REF
QH
9
.S8
1977   Linne, Carl von,
Cop.1   1707-1778.
        Miscellaneous tracts
        relating to natural
        history, husbandry,
        and physick

THE CHICAGO PUBLIC LIBRARY

FOR REFERENCE USE ONLY
Not to be taken from this building

REF
QH
9
.S8
1977
Cop.1

The Chicago Public Library

Received _____ FEB 24 1979 _____

© THE BAKER & TAYLOR CO.

MISCELLANEOUS TRACTS
RELATING TO NATURAL HISTORY,
HUSBANDRY, AND PHYSICK

This is a volume in the Arno Press collection

HISTORY OF ECOLOGY

Advisory Editor
Frank N. Egerton III

Editorial Board
John F. Lussenhop
Robert P. McIntosh

*See last pages of this volume for a
complete list of titles.*

MISCELLANEOUS TRACTS
RELATING TO
NATURAL HISTORY
HUSBANDRY
AND
PHYSICK

Carl Linnaeus

Translated by Benjamin Stillingfleet

ARNO PRESS
A New York Times Company
New York / 1977

Editorial Supervision: LUCILLE MAIORCA

Reprint Edition 1977 by Arno Press Inc.

HISTORY OF ECOLOGY
ISBN for complete set: 0-405-10369-7
See last pages of this volume for titles.

Manufactured in the United States of America

Library of Congress Cataloging in Publication Data

Linné, Carl von, 1707-1778.
   Miscellaneous tracts relating to natural history, husbandry, and physick.

   (History of ecology)
   Reprint of the 1791 ed. published by J. Dodsley, London.
   Includes index.
   1. Natural history—Collected works. 2. Ecology—Collected works. I. Stillingfleet, Benjamin, 1702-1771. II. Title. III. Series.
QH9.S8 1977   574.5'08   77-74237
ISBN 0-405-10406-5

MISCELLANEOUS TRACTS

RELATING TO

NATURAL HISTORY, HUSBANDRY, AND PHYSICK.

To which is added the

CALENDAR of FLORA.

By BENJ. STILLINGFLEET.

THE THIRD EDITION.

LONDON:

Printed for J. DODSLEY, in Pall-Mall; BAKER
and LEIGH, in York-Street, Covent-Garden;
and T. PAYNE, at the Mews Gate.

MDCCLXXV.

Homo naturæ minister et interpres tantum facit et intelligit, quantum de naturæ ordine, re vel mente obfervaverit : nec amplius fcit vel poteft. BACON.

Primus gradus fapientiæ eft res ipfas noffe ; quæ notitia confiftit in vera idea objectorum; objecta diftinguuntur et nofcuntur ex methodica illorum *divifione* et conveniente *denominatione*; adeoque divifio et denominatio fundamentum noftræ fcientiæ erit. LINN.

TO THE
RIGHT HONORABLE
George, Lord Lyttelton,
Baron of FRANKLEY.

My Lord,

BESIDES private motives of respect and honor, there is another of a public nature, which makes me desirous to inscribe the following sheets to Your Lordship. I mean the zeal which You shewed in Parliament for securing to this Country that noble collection of natural curiosities now reposited in the British Museum; which cannot

DEDICATION.

fail in time to produce many good effects, and prove the truth of what Your Lordship observed, that the reputation and interest of the nation were highly concerned in that purchase. I am, with the greatest regard,

MY LORD,

Your Lordship's most Obliged

and Humble Servant,

BENJ. STILLINGFLEET.

PREFACE

OF THE

TRANSLATOR.

THE following pieces were selected from many others published by several ingenious members of that great, and hitherto unrivalled school of natural history, the university of Upsal in Sweden, under the presidency of Linnæus [a]. They were

---

[a] Linnæus Phil. Bot. p. 9. has these words, vid. dissert. nostra de ficu. Now there is a piece in Amœn. Acad. vol. 1. on this subject, in which the matter referred to is contained. This piece goes under the name of Cornelius Hegardt, tho' Linnæus plainly quotes it as his own.

How

were selected not as the only, or even the most valuable, but as answering best the intention of the translator; which was to make known more generally how far all mankind is concerned in the study of natural history, and thereby to incite such as are properly qualified to prosecute, and encourage that branch of knowledge, and spread, as far as the nature of the thing is capable of, amongst all orders of men in this nation, the improvements made in it by the excellent Linnæus. His name, it must be confessed, has been for some time past in the mouths of people, but his works, i imagine, are little known except to a few vertuosi who have a more than ordinary curiosity, and ardor to look into the minute parts of nature. It cannot indeed be otherwise. For to understand him and to make use of his method, requires

---

How far that may be the case of all the other pieces in the Amæn. Acad. i cannot pretend to say. But it is most likely from the practice in foreign universities in relation to these held for degrees, that they must in great part be attributed to him, as president.

PREFACE.

more patience and time than are likely to fall to the share of the generality of the world. My design therefore is not to exhort people indiscriminately to study his works; but, as i observed before, to give them some idea how usefull his pursuits are likely to become in many respects. There will appear, i imagine, such great and extensive views in relation to husbandry, physic, and the general œconomy of human life in the few specimens i have given, that in this age and nation, where every art and science, that can be of any use to the public, are sure to meet with generous encouragers, there will be found many who will readily promote any rational endeavour to push these discoveries farther, or put in practice such hints, as may seem to bear a probable appearance of success.

I can scarcely condemn mankind for treating with contempt a vertuoso whom they see employed in poring over a moss or an insect day after day, and spending his life in such seemingly unimportant and barren speculations. The first and most natural

ral reflections that will arise on this occasion must be to the disadvantage of such pursuits. Yet were the whole scene of nature laid open to our views, were we admitted to behold the connections and dependencies of every thing on every other, and to trace the œconomy of nature thro' the smaller as well as greater parts of this globe, we might perhaps be obliged to own we were mistaken; that the Supreme Architect had contrived his works in such a manner, that we cannot properly be said to be unconcerned in any one of them; and therefore that studies which seem upon a slight view to be quite useless, may in the end appear to be of no small importance to mankind. Nay, were we only to look back into the history of arts and sciences, we must be convinced, that we are apt to judge over hastily of things of this nature. We should there find many proofs, that he who gave this instinctive curiosity to some of his creatures, gave it for good and great purposes, and that he rewards with usefull discoveries all these minute researches.

PREFACE. ix

It is true this does not always happen to the searcher, or his contemporaries, nor even, sometimes to the immediate succeeding generation; but i am apt to think that advantages of one kind or other always accrue to mankind from such pursuits. Some men are born to observe and record what perhaps by itself is perfectly useless, but yet of great importance, to another who follows and goes a step farther still as useless. To him another succeeds, and thus by degrees, till at last one of a superior genius comes, who laying all that has been done before his time together, brings on a new face of things, improves, adorns, exalts human society.

All those speculations concerning lines and numbers so ardently pursued, and so exquisitely conducted by the Grecians; what did they aim at? or what did they produce for ages; A little arithmetic, and the first elements of geometry were all they had need of This Plato asserts, and tho' as being himself an able mathematician and remarkably fond of these sciences, he
recom-

recommends the study of them, yet he makes use of motives that have no relation to the common purposes of life.

When Kepler, from a blind and strong impulse merely to find analogies in nature, discovered that famous one between the distances of the several planets from the sun, and the periods in which they compleat their revolutions; of what importance was it to him or to the world?

Again; when Galileo, pushed on by the same irresistible curiosity, found out the law by which bodies fall to the earth, did he or could he foresee that any good would come from his ingenious theorems, or was any immediate use made of them?

Yet had not the Greeks pushed their abstract speculations so far; had not Kepler and Galileo made the above-mentioned discoveries; we never could have seen the greatest work that ever came from the hands of man. Every one will guess that i mean Sir Isaac Newton's Principia.

Some obscure person, whose name is not so much as known, diverting himself idly

as

PREFACE.　　xi

as a stander-by would have thought, with trying experiments on a seemingly contemptible piece of stone, found out a guide for mariners on the ocean, and such a guide as no science, however subtile and sublime its speculations may be, however wonderful its conclusions, would ever have arrived at. It was bare curiosity that put Sir Thomas Millington upon examining the minute parts of flowers; but his discoveries have produced the most perfect, and most usefull system of botany that the world has yet seen.

Other instances might be produced to prove, that bare curiosity in one age is the source of the greatest utility in another. And what has frequently been said of chymists, may be applyed to every other kind of vertuosi. They hunt perhaps after chimæras and impossibilities, they find something really valuable by the bye. We are but instruments under the Supreme Director, and do not so much as know in many cases what is of most importance for us to search after. But we may be sure

sure of one thing, viz. that if we study and follow nature, whatever paths we are led into, we shall at last arrive at something valuable to ourselves and others, but of what kind we must be content to remain ignorant.

I am sensible that after all i have said, or can say, many people will not be persuaded to allow that the study of some parts of natural history can be worthy of a rational creature. They will never vouchsafe to look on mosses and insects in this light. Yet why may not the study of these likewise have its use in future times? It ought to be considered that the number of the latter is immense, that it is but lately that any great attention has been paid to them, that one of them is and has been long the means of cloathing thousands and feeding more, that another affords us honey, another a fine dye, not to mention some few besides, of acknowledged benefit to mankind. Lastly, that they are capable of doing us the greatest mischief, and that it is possible that a more thorough knowledge of them

them may inſtruct us how to ſecure ourſelves againſt their attacks. Whether this be poſſible we can never know, till proper encouragement has been given to this branch of natural hiſtory. Something to the ſame purpoſe might be ſaid concerning moſſes, but as the intent of one of the following pieces is principally to take off ſuch objections as i have been conſidering, i ſhall dwell no longer on this ſubject; but proceed to give a ſhort account of what Linnæus has done towards the improvement of natural hiſtory, that the reader, who is unacquainted with his works, may form ſome idea, tho' very imperfect, of this great man. Firſt then, he has invented a new ſyſtem of botany, founded on the male and female organs of generation in plants, a ſyſtem which has thrown a new light over botany. He has defined about 10,000 plants, ranged them into claſſes, genera and ſpecies, given new and regular generical names to many inſtead of thoſe barbarous and uncouth ones which prevailed till his time, and added ſpecifical names

PREFACE.

names to all, short, easy, and oftentimes significant, a thing never so much as attempted before. He has brought into botany, a precision, concisenefs, and elegance, that were very much wanted. He has observed and given names to some parts of plants not taken notice of by any other botanist, parts which in some cases are sufficient as well as necessary to distinguish the genus and the species.

The Philosophia botanica [b] of this author affords throughout instances of this reformation. Had he wrote no other book but this, he would have deserved the highest praise from all lovers of botany. For besides the improvements just mentioned it comprehends in a short compass some-

---

[b] In the year 1750, when he was writing this book, as he tells us in the preface to it, he was hindered from going on by a terrible fit of the gout, that broke the strength of his mind as well as body. In the year 1755, he says, Flor. Suec. article 450, that he had been freed from the gout for some years by eating great quantities of fresh strawberries. He adds that this fruit dissolves the tartar of the teeth, that it is remarkably good for people afflicted with the stone or gout, and that it may be safely eaten in abundance.

PREFACE.　　xv

thing of confequence in every branch of that part of natural hiftory, and affords hints for various difcoveries, hints that muft, if purfued, produce many confiderable improvements in phyfic, hufbandry, and œconomy.

He has publifhed a Materia medica fo far as relates to plants, in which he has undertaken to determine many fpecies commonly ufed, but not fufficiently afcertained, adding throughout in the fhorteft manner poffible what he has found to be ufelefs or efficacious, and as he affures us never highly recommends any without being thoroughly convinced of their vertues by his own experience in the hofpitals where he prefided. Some of thefe medicines have not yet, i believe, been received into our fhops, but they may poffibly deferve confideration.

In the laft edition of his Syftema naturæ he has mentioned above 1500 fpecies of infects, has claffed them all, divided them into genera and fpecies, defcribed them as to the minuteft parts fo far as was neceffary to diftinguifh them, marked the places where

where they are to be found, the plants they feed upon, their transformations, cited the authors who have treated on them, given them claffical, generical, and trivial, or fpecifical names; has done the fame by birds, fifhes, and all other known animals; has ranged all the foffils, minerals and ftones, to ufe his language, in a manner partly borrowed, and partly founded on his own obfervations. But what improvements and additions he has lately made to this part of natural hiftory, as well as that of plants, we cannot fay till the other part of his new edition of the Syftem of nature comes out, which is expected daily. However what we fee he has done in relation to animals, leaves us no room to doubt but that it will altogether be the moft extraordinary book that was ever publifhed in this or almoft any other way.

Befides his writings, of which i have mentioned but a fmall part, this indefatigable man, born to be nature's hiftorian, has travelled over Lapland, all Sweden, part of Norway, Denmark, Germany, Holland,

PREFACE. xvii

Holland, England, and France, in search of knowledge. That part of his travels which is published in Latin has many curious and useful observations relating to the purposes of common life. Of those which are written in his own tongue i cannot give any other account, but that by some quotations from them to be found in the writings of his disciples it appears, that they very well deserve to be communicated to the world in a language more generally understood.

Besides these labours of his own, the world will be one day obliged to him for what others have done. Incited by his example and persuasion, C. Ternstrom went into Asia; P. Kalmius to Pensilvania and Canada; L. Montin into one part of Lapland; D. Selander into another; F. Hasselquist into Ægypt and Palestine; O. Toren to Malabar and Surat; P. Osbech to China and Java; P. Loefling to Spain and America; P. J. Berg to Gothland; M. Koehler to Italy and Apulia; and D. Rolander to Surinam and St. Eustacia; all these with a view to the promotion of natural history. When we consider him in this light of a master of such disciples as these, and many others, some of whose works make up the following book, he must appear like Homer at the head of the poets, Socrates at the head of Greek moralists, and our Newton at the head of the mathematical philosophers. Among all these extraordinary qualifications there appear throughout his writings spirit, candor, a due regard for others, and proper modesty and diffidence of himself.

I will give a short specimen of his way of thinking in relation to the degree of human knowledge hitherto attained by man on the subject of natural history. A subject on which it was very natural for a less extensive genius to be vain, as he has had so great a share himself in the advancement of it. The passage is taken out of the introduction to the new edition of his System of nature, and is to this effect. 'How 'small a part of the great works of nature 'is laid open to our eyes, and how many

PREFACE. xix

'things are going on in secret which we
'know nothing of! How many things are
'there which this age first was acquainted
'with! How many things that we are ig-
'norant of will come to light when all
'memory of us shall be no more! For
'nature does not at once reveal all her se-
'crets. We are apt to look on ourselves as
'already admitted into the sanctuary of her
'temple, we are still only in the porch. I
'have entered, adds he, into the thick and
'shady woods of nature, which are every
'where beset with thorns and briars. I
'have endeavored as much as possible to
'keep clear of them, but experience has
'taught me that there is no man so cir-
'cumspect as never to forget himself, and
'therefore i have born with patience the
'sneers of the malevolent, and the buffoo-
'neries of those whose vivacity is exerted
'only to molest and give offence to others.
'I have, in spite of these insults, kept on
'steadily in my old path, and have finished
'the course i was destined for.'

The

The latter part of this passage, shews that he has not been without his enemies, and that he hath suffered in the same way that all the most curious enquirers into nature have done in all ages. The tartness of his expressions, which is still stronger in the original, plainly proves that they have not used fair arguments against him, but like interested rivals, or men of a superficial understanding, have endeavored to subject him and his labors to ridicule. But whatever has been his fate in his own countrey, as far as i know, his name is almost universally mentioned with respect in all other parts of Europe. It is true, objections have been made to his innovations in other places besides Sweden [c],

which

Having since the first edition of these tracts met with Browellius' answer to Siegesbec, M. D. and botanical professor at Peterburg; who wrote against the sexual system of Linnæus, i cannot omit quoting one of his objections, which i imagine will divert the reader, at the same time that it may serve as an instance how far zeal for old notions will sometimes carry men. The objection is, that the laws of nature are overturned by

Linnæus,

PREFACE.

which muſt unavoidably happen on many accounts, but particularly becauſe thoſe natural hiſtorians who had been brought up and inured to other ſyſtems, who

Linnæus, ſince polygamy and adultery would be according to his ſyſtem allowed in the vegetable world; for in ſome plants there are many filaments to one piſtil. This is polygamy. In others there are female flowers, which are impregnated by the duſt of male flowers, which have other female flowers belonging to them, i. e. which are already married. This is plainly adultery. Now according to profeſſor Siegeſbec, it is not credible that ſuch confuſion and deteſtable pollution ſhould be tolerated in nature.

Browellius rightly obſerves in his anſwer, that Siegeſbec had totally overlooked many inſtances of theſe enormities in the animal kingdom, and even the immorality of farmers and their wives, ſhepherds, jockies, ſportſmen, nay even ladies of reputation, who in their ways promote theſe immoral and indecent practices.

However it muſt be obſerved in favor of this very ſcrupulous profeſſor, that ſyſtems of philoſophy, founded on facts, have been anathematiſed, and the authors and favorers of them condemned to the ſevereſt puniſhments, for reaſons as little to the purpoſe as the foregoing of Siegeſbec. To quote inſtances would be endleſs, as every one the leaſt converſant in the hiſtory of learning will eaſily recollect them. But ſo moderate is the world now become, that i do not hear that the Linnæan ſyſtem is looked on as heretical even at the court of Rome, though the profeſſor has drawn ſome ſhrewd arguments againſt it from the book of Geneſis.

PREFACE.

had learned things by other names, and could not eafily attain the new ones, muſt have ſtrong prejudices ariſe on this occaſion. This objection being perſonal i ſhall not conſider it any farther, but readily allow that great indulgence is due to ſuch people, and that their fate is to be pitied for coming into the world too ſoon to be enlightened farther on ſubjects, that perhaps had employed the greateſt part of their life. But there are prejudices of another ſort which i cannot omit to conſider more fully on this occaſion.

In order to this it muſt be premiſed, that the uſe and intent of a claſſical ſyſtem in any part of natural hiſtory, is not to range things according to their natural connections in regard to their outward aſpects, or eſſential qualities, or their medicinal or œconomical properties, but to range them in ſuch a manner that upon a plant, mineral or animal being ſhewn to a naturaliſt, he may certainly, upon a due inſpection of the object, give its true name according to ſome ſyſtem. He who goes

farther

farther than this is not barely a naturalist, but something more, viz. a physician, a chymist, a farmer, a gardener, &c. And he who cannot go thus far to a certain degree, does not deserve the name of a naturalist, however skillfull he may be in the vertues and properties of bodies animate and inanimate.

The use then and intent of a classical system is nothing more than that of a dictionary, where no one complains that words totally unconnected in sense are put near one another. The question therefore as to the sexual system [d], v. g. in plants, is not whether they be ranged naturally, but whether in the best manner possible in order to be known. Nay farther, it matters not whether the sexual system be founded on nature or not, i. e. whether there be any

---

[d] At the end of the preface i have endeavored to explain the meaning of these terms in such a manner, that i think any curious person that will be at the pains to compare my explication with nature, cannot fail to understand perfectly what they mean in general. I thought this method would be more agreeable to the reader than to be referred to other books.

propagation by seeds without male and female organs of generation. The whole to be considered is whether those parts which are called, and, i believe truly called so, do really exist, and whether they for the most part exist so uniformly, as to furnish marks sufficient to distinguish the classes, &c. by. Nor does it matter whether it be hard to distinguish those marks, but whether they can with proper care and patience be distinguished, and whether we can surely distinguish plants, without observing those nice and minute parts, and whether a system has been found equally sure with the Linnæan without having regard to those parts. Those who think so would do well to inform the world of their discovery, and not make objections that affect only the obscurity of nature, when they mean to condemn a system which is obscure merely from its consonancy to nature. If Providence has thought fit to write in cyphers, shall he be blamed who endeavors to give a key to its works, because
some

some men cannot distinguish one stroke from another in the cypher?

Those who have not learned to read the characters of nature for want of leisure, patience, or any other cause, ought not to complain that Linnæus cannot make them skillfull in a part of knowledge they are not qualified for. If a man unacquainted with the learned languages wants to know the meaning of a Greek word, will he complain of the lexicon, because he cannot find it? certainly not. Neither ought we to complain of Linnæus in a similar case.

This i think is a full answer to all the objections that have or can be made to his system in general. What errors he has committed according to his own principles in relation to particulars is quite another question. I am one of those who think him not free from errors. Nor is it wonderfull that he should fall into some, but it is truly wonderfull that one man should be able to invent and carry so far so nice and extensive a system, especially when we consider not only what he has done in botany, but what he has done in all the branches of

natural

natural history besides, and some of them almost entirely neglected before his time. I should therefore wish that those who are fond of this part of knowledge would, instead of making frivolous objections, try by an accurate and diligent examination to rectify his mistakes, and thereby help to perfect a system which deserves the utmost attention, and commendation.

Tho' i said above that it matters not whether the sexual system be founded on nature or not; yet it was natural for the inventor of it to endeavor by all proper means to vindicate it as likely to be so, and this he has done to the satisfaction of the most curious observers; and i will venture to add that it is natural for others likewise to embrace with zeal a system, that puts the works of Providence in so new and beautiful a light by continuing the analogy from the animate to the inanimate creation. It seems as if Providence intended to lead men to this discovery by striking our senses so intensely and so agreeably with those very parts which contain the clue of this system. Yet such is the inattention and in-

PREFACE.

inaccuracy of man on certain points, that even a tolerable conjecture concerning the use of those parts was not made till the year 1676.

Having finished all that i think necessary to say concerning Linnæus and his works, i shall now come to what relates immediately to myself only. First then as to the translation, i have endeavoured to avoid making it too literal, and servile, but yet i hope without taking any undue liberties, or deviating from the sense of the originals.

The part which is likely to prove least agreeable to the reader, is that which was most troublesome to the translator. I mean the names of things not generally known. Some of these i have been obliged to leave in Latin, not being able to find any English names for them. I will not pretend to have avoided all mistakes on this head, but it is certain i should have committed more, as well as have had much more trouble, had it not been for the assistance of the ingenious Mr. Hudson, whose skill in all the branches of natural history, and particularly

cularly those relating to his profession as an apothecary, cannot fail to recommend him to the favour of the public. To him i likewise owe the ascertaining of some of the grasses, one of which, viz. the small bent grass which i had in my collection, but knew not where i found it, he discovered to be the gramen minimum anglo-britanum, mentioned in the indiculus plantarum dubiarum at the end of Ray's Synopsis.

I must not omit also on this occasion to acknowledge my obligation to that excellent botanist Dr. Watson, for favoring me with a perusal of his collection of grasses, which was of no small service to me.

But to return to the translation; I said that i did not pretend to have avoided all mistakes in relation to the names of things, i will now extend this farther, and own my suspicions that i may have made some in relation to other particulars, but i hope they are of such a kind only as may be looked on with indulgence by the learned, especially when they consider the great variety of subjects

PREFACE.

jects treated on in these pages, of none of which subjects i profess to be a master, and therefore do not undertake to teach such readers; but on the contrary shall always be ready and even desirous to receive instructions from them. I beg they will also consider that i do not aim at letting the unlearned into the mysteries of this part of knowledge, or even teaching them the elements of it. My business is only to excite curiosity, and therefore small errors can be of no consequence. What i have farther to say will be found in notes.

Res summas initio deberi parvo ac debili experientia omnium temporum testatur. Amænit. Acad. vol. 2. p. 266. §. 2.

End of the PREFACE.

IN order to explain the fexual fyftem, i fhall make ufe of the lilly, as that plant is almoft every where to be found, and as the parts of generation are in that more obvious, than perhaps in any other flower. Upon opening the flower leaves there will appear in the very center, at the bottom, an oblong thickifh fubftance with fix furrows along its fides. This contains the feeds, and is called

The germen or germ.

On this ftands a fmall kind of pillar called

The ftyle.

Which is terminated by a thickifh triangular head, called

The ftigma.

Thefe all together form the female part of the flower, and are called by one name,

The piftil.

Round this piftil grow fix long thready fubftances, called

The filaments,

Each terminated by an oblong body, that plays as on a pivot, upon the leaft motion

tion being given to the flower, and is called
The anthera.

This anthera contains the male duſt, which when ripe is ſcattered about by every breath of air, and what happens to fall on the ſtigma, or upper part of the piſtil, is ſuppoſed to enter thro' the ſtyle into the germ, and there impregnate the ſeed.

This plant is called an hermaphrodite, becauſe the male and female organs of generation are contained within one flower. Moſt plants are hermaphrodites, like this, and have ſomething analogous to what i have deſcribed above. Some plants have the male and female parts ſeparate on the ſame individual; others have male parts on one plant and female on another.

The part of the flower that contains honey, is called

The nectary.

Only a few plants have this part, the lilly has it; but as the knowledge of it is not neceſſary for underſtanding the following pieces, i ſhall not trouble the reader with a deſcription of it.

CON-

CONTENTS.

An oration concerning travelling in one's own countrey, by Dr. Linnæus —— —— page 1

The œconomy of nature, by Isaac Biberg —— —— —— 37

On the foliation of trees, by Harald Barck —— —— —— 131

Of the use of curiosity, by Christoph. Gedner —— —— —— 159

Obstacles to the improvement of physic, by J. G. Beyerstein —— 201

The calendar of Flora —— —— 229

The Swedish Pan, by Nicholas Hasselgren —— —— —— 339

Observations on grasses, by the Translator —— —— —— 363

BENEFIT

OF

TRAVELLING, &c.

Miscellaneous Tracts, &c.

An ORATION concerning the necessity of travelling in one's own countrey, made by Dr. LINNÆUS at Upsal, Oct. 17, anno 1741, when he was admitted to the royal and ordinary profession of physic.

Amænitat. Academ. vol. ii.

MOST honorable and most learned auditors of all orders, i am going to undertake a province allotted to me by the favor of our most august, and most potent monarch, whose will it is that i preside over, and direct the study of physic in this University; and that i do my utmost to advance the glory of this illustrious body. May his choice be crowned with success, and may the great and good God favor my undertaking.

As by custom, delivered down by our forefathers, and prescribed by the laws of our academy,

demy, i am obliged upon undertaking this province to say something before so illustrious a circle of fathers and citizens; i confess that all those circumstances, each of which is apt to strike terror into the mind of man, offer themselves together in a croud before my eyes on this occasion. For whether i consider the ampleness of the place, or the dignity of the audience, or the multitude of chosen people, or lastly my little talents in the arts of speech; all these circumstances, i ingenuously confess, throw me into no small confusion.

For if the most eloquent men, when they come to speak in public, have been known to tremble, and become incapable of uttering a single word; what must i feel who have none of the common advantages, either from art or nature, in the readiness and elegance of speech?

However, since I am under a necessity of saying something, i must fly for refuge to that favor, and humanity, which you never refuse to those who speak on these occasions; and thus i doubt not but that, however deficient i may be from want of talents, or want of exercise, i shall not wholly fail of the end i aim at. I shall therefore, most honorable auditors, undertake to treat on a subject neither unsuitable to

the

TRAVELLING, &c.

the present occasion, nor to the office i am going to enter into, nor to that employment which i was lately engaged in by the will, and suffrage of the high, and mighty states of this kingdom; and from which i am now once again brought back to this seat of the muses. Nay so far is the subject i am about to treat on, from being unsuitable to any of these circumstances, that it seems to me particularly adapted to every one of them. The subject is concerning the necessity of travelling in one's own countrey, and the advantages that may thence accrue, especially to physicians. I shall treat it in a plain and popular manner; and endeavor to manage it so, that the meanness of my language may be compensated by the dignity of the matter, and the brevity of my expressions.

All human knowledge is built on two foundations; reason and experience.—These two joyned together are necessary to make a good physician.

We must confess indeed, that the business of reasoning may be carried on with equal success in our closets, as in travelling, supposing we have an opportunity of conversing with men truly learned.

But it was experience, that sovereign mistress without which a physician ought to be ashamed to open his lips; it was experience, i say, that consecrated to immortality so many of the antients, and amongst the rest that divine old man Hippocrates, whose writings were published many ages before christianity. The writings of this wonderful man alone, among so many ingeniously contrived systems, remain to this day, and will for ever remain firm, unmoved, unshaken, untouched by any decay, by any change. It is experience that has adorned with laurels the heads of so many celebrated physicians in all times, and even now adorns. And hence it is that the chief and most honorable title of physician is to be called a man of experience. Experience ought to go first; reasoning should follow. The former furnishes the materials of knowledge; the latter holds her consultations on the given phænomena; and when she has weighed with judgment every circumstance, she discovers truth, and concludes, orders, and determines rightly about the point in question. Experience ought to be animated by reason in all physical affairs; without this she is void of order, void of energy, void of life. On the other hand reason without experience can do nothing; being nothing, but the mere dreams, phantasms, and meteors of ingenious men who abuse their time. The antients certainly did not, any more than we, bring experience into the world with them. There is need of much diligence and labor, before man can be thoroughly instructed. Dioscorides confesses, that he undertook many journies in order to increase experience; and the other fathers of physic in their writings frequently make mention of their travels either expressly or tacitly.

Academies were instituted to the end, that men well versed in all kinds of literature, and enriched besides by much experience, might be invited thither, and that the youth, who were ambitious of becoming learned, might flock together to those seats; and have the advantage of improving no less by the experience, than by the erudition of the professors; and these qualifications combined together, which is of all alliances the most pleasing, very justly deserve the utmost veneration and respect.

Vast and sumptuous libraries are erected in academies; in which the observations of the learned, like so many legacies, and donations, are preserved; that they who diligently give themselves up to study, may become endued with learning, polished, and confirmed by experience. These libraries are the repositories of wisdom, and their stores are laid open to every ingenious candidate.

Hospitals are founded that the candidates of physic may learn those things at the patients bed side, which cannot be learned from books; for here practice, and experience shew their force by means of the eyes, and hands; as he paints any object most accurately, who paints from the idea, which his own eyes afford him, and not from that, which he gets by the relation of another.

Anatomy schools are erected, that we may behold in another's body, as it were in a glass, the nature, and constitution of our own; as those conceive more clearly the situation of countries, districts, and cities, and the manners, rites, and customs of their inhabitants, who themselves have been there, and have seen what is remarkable amongst them with their own eyes, than he who relies solely upon the vague and imperfect maps, and relations of travellers.

Phyſic gardens are here cultivated; where the plants of various kinds are collected from all parts of the globe, that we may by this means behold, as it were, the great in the little world.

Hither inſtruments for experimental philoſophy are brought together, that the abſtruſe forces of the elements, which otherwiſe would eſcape our ſenſes, may be made manifeſt, and that ſo we may ſucceſsfully be let into the very receſſes of nature; as far as human penetration will admit of.

All theſe things are inſtituted in academies, that the youth may arrive at knowledge by experience; all tend to this end, that tho' we be confined to one ſpot, one corner of the earth, we may examine the great and various ſtores of knowledge, and therein behold the immenſe domains of nature, and get acquainted with ſuch things, as otherwiſe muſt be ſought for, and oftentimes in vain, over the whole globe.

In my opinion therefore ſtudying at academies ought by no means to be neglected, but rather ſhould be looked upon as neceſſary to thoſe, who are ambitious of attaining wiſdom, ſupported by experience. And thoſe who endeavor

deavor to inftill into the minds of young people a contempt for univerfities, and to withdraw the ftudious from thefe feats of learning, fuggeft very pernicious advice; not confidering that in thefe ftorehoufes of knowledge much greater, and more excellent things may be attained by means of experience in a very fhort fpace of time, than by the moft multifarious, moft indefatigable, and moft extenfive reading at home all one's life.

If i may be allowed to fpeak what is really fact, this our univerfity may contend with any foreign one whatever for true, and folid learning in all thofe parts of knowledge, which i have enumerated, owing to our noble, and exemplary inftitutions. For we begin to excell in botanical gardens, in hofpitals, in apparatus's for experimental philofophy, in anatomical preparations, and other helps for arts and fciences, and to excell fo much that we are likely in time, by the blefling of the almighty, to be inferior to no univerfity.

Although fome univerfities excell others on account of certain advantages peculiar to themfelves; for in proportion as one kind of knowledge in this, or that nation is held in greater, or lefs efteem, and is therefore

more

more or less cultivated, so the professors of it will be more or less skillful; as at this time the hospitals at London both for number and goodness exceed all others, at Paris chirurgical operations, at Leyden anatomical preparations, at Oxford botanical collections; tho', i say, this may be the case, yet i cannot think, that those act prudently, or enough consult the good of themselves, and countrey, who seek for that abroad, which may be had at home, and who travel to foreign universities, before they have laid a sufficient foundation in their own countrey. And there is no doubt but that they who do so will at last repent of their error. He, who goes abroad raw, and ignorant, seldom returns more learned. Whereas, he, who has spent his time well at his own university, will never find reason to repent. Whoever has employed himself properly in the study of the arts, and sciences will become an usefull, and solid man in every branch of business. Whoever, before he sets out to visit regions warmed by other suns, has laid the first foundations of his studies in his native countrey, will be most likely to bring back materials of far greater price than we usually see amongst the greatest part of our tra-

travellers, who seldom return home laden with any thing, but fine sounding, and empty words collected out of the European languages. What do they learn, but to prate about theatres, and plays, and the modes of dress amongst the Italians, the Spaniards, the Germans, and above all the French? If they were well advised they would not stir a foot out of their own countrey; that they might not destroy their fortunes, their time, their health, nay their very life itself by luxury, and voluptuousness. They would not then return, as too frequently happens, entirely useless to themselves, and countrey, and a burthen upon the face of the earth. But whither am i hurried?

My design was, in the little time allotted me, to speak to you, gentlemen, not of the peculiar advantages of universities, or of sojourning at this, rather than any foreign one; but chiefly of travelling in one's own countrey, thro' its fields and roads; a kind of travelling, I confess, hitherto little used, and which is looked upon as fit only for amusement. I once more, most honorable auditors, beg your patience, and that i may not forfeit all right to your favor, and benevolence, i promise to be as short as possible. You know what the poet says,

The

TRAVELLING, &c.

The farmer talks of grasses and of grain,
The sailor tells you stories of the main.

You ought not therefore to wonder, that i choose to make travelling in one's own countrey the subject of my discourse. Every one thinks well of what belongs to himself, and every one has pleasures peculiar to himself. I have travelled about, and passed over on foot the frosty mountains of Lapland, have climbed up the craggy ridges of Norland, and wandered along its steep hills, and almost impenetrable woods. I have made large excursions into the forests of Dalecarlia, the groves of Gothland, the heaths of Smoland, and the unbounded plains of Scania. There is scarcely any considerable province of Sweden, which i have not crawled through and examined; not without great fatigue of body and mind. My journey to Lapland was indeed an undertaking of immense labor; and i must confess, that i was forced to undergo more labor, and danger in travelling thro' this one tract of the northern world, than thro' all those foreign countreys put together, which i have ever visited; though even these have cost me no small pains, and have not a little exhausted my vigor. But love to truth, and gratitude towards the su-

preme

preme being oblige me to confefs, that no sooner were my travels finifhed, but, as it were, a Lethæan oblivion of all the dangers, and difficulties came upon me, being rewarded by the ineftimable advantages, which i reaped from thofe devious purfuits. Advantages, the more confpicuous for that i became daily more and more fkillful, and gained a degree of experience, which i hope will be of ufe to myfelf, and others; and, what i efteem above all other confiderations; as it comprehends in one all other duties, and charities; to my countrey, and the public.

Good God! how many, ignorant of their own countrey, run eagerly into forreign regions, to fearch out and admire whatever curiofities are to be found; many of which are much inferior to thofe, which offer themfelves to our eyes at home. I have yet beheld no forreign land, that abounds more with natural curiofities of all kinds, than our own. None which prefents fo many, fo great, fo wonderfull works of nature; whether we confider the magazines of fnow heaped up for fo many ages upon our Alps, and amongft thefe vaft tracks of fnow green meadows, and delicious vallies here and there peeping forth, or the lofty heads

TRAVELLING, &c.

heads of mountains, the craggy precipices of rocks, or the sun lying concealed from our eyes for so many months, and thence a thick Cimmerian darkness spread over our hemisphere, or else at another season darting his rays continually along the horizon. The like to all which in kind, and degree, neither Holland, nor France, nor Britain, nor Germany, nor lastly any countrey in Europe can shew; yet thither our youth greedy of novelty flock in troops. But it was not my intent to speak of these things at present. I come now closer to my purpose, being about to shew by instances, that the natural philosopher, the mineralogist, the botanist, the zoologist, the physician, the farmer, and all others, initiated in any part of natural knowledge, may find in travelling thro' our own countrey things, which they will own they never dreamed of before. Nay things which to this day were never discovered by any person whatever. Lastly, such things, as may not only gratify, and satiate their curiosity; but may be of service to themselves, their countrey, and all the world.

To give a few examples. The sagacious searcher after nature will find here wherewithall to sharpen, and exercise his attention in beholding

holding the top of mount Swucku, of so immense a height, that it reaches above the clouds. The wonderfull structure of mount Torsburg, the horrid precipices of the rock Blakulla in an island of that name, situated near Oeland, and that presents by its name, still used among the Suegothic vulgar, no less than by its dismal aspect, an idea of the stupidity, and superstition of that antient people.

Besides the wonderfull vaults and caverns of the Skiula mountains, the high plains of the island Carolina, the unusual form and structure of the Kierkersian fountains in Oeland; to pass over numberless other strange works of nature, the like to which perhaps are no where to be met with.

Where can we have greater opportunities, than in this Suegothic tract, of considering the intense rigor, and vehemence of winter, the incredible marble-like strength of ice? And yet in this inclement climate grain of all sorts is observed to spring forth sooner, grow quicker, and ripen in less time than in any other part of the world.[a]

[a] Vid. a treatise concerning the foliation of trees published in this collection, and the prolegomena to the Flora Lapponica of this author, where he says that at Purkyaur in Lapland anno 1732. barley sown May 31. was ripe July 28. i. e. in 58 days; and rye sown May 31. was ripe, and cut Aug. 5. i. e. in 66 days.

Whoever desires to contemplate the stupendous metamorphoses of sea, and land, will scarcely find any where a more convenient opportunity, than in the south, and east parts of Gothland; where the rock-giants, as they are called, seem to threaten heaven, and where the epocha's of time, the ages, the years, if i may so say, are as it were carved out in a surprising series upon the sea-shore, and the ground above the shore.

The philosopher will find room to exercise his ingenuity sufficiently in the Oeland-stone, by trying to discover how to overcome its moist nature, and quality, which whoever could accomplish would do no small service to his countrey, and above all would infinitely oblige the inhabitants of that place.

I shall say no more than what is known, and confessed by all the world, when i say that there is no countrey in the habitable part of the globe, where the mineralogist may make greater progress in his art, than in this our countrey. Let any one, that can, tell me, in what regions, more rich, and ample mines of metal are found, than in Sweden, and where they dig deeper into the bowels of the earth than here.

Let the mines of Norburg, the ridge of Taburga, the pits of Dannemore, Bitſberg, Grengia, and laſtly the immenſe treaſures of Salbergen, and Fahluna be my witneſs, which exceed all in the known world.

Where do the poſſeſſors ſuffer forreigners more freely to approach their furnaces, and obſerve their operations? where are there men more ready to communicate their knowledge? Strangers are received by us with civility, and even preſſed to ſtay.

Who would not ſhudder on beholding thoſe forges, vomiting forth immenſe clouds of fire and ſmoak, where our iron ores are melted? who would not behold with pleaſure the ſimple countreyman in the thick pine-groves of Dalecarlia, without furnace, without any apparatus, extracting an iron ſo very fit for uſe, that it yields to no other, tho' prepared with the fierceſt fires, and greateſt expence?

Who ten years ago would have imagined, that the *lapis calaminaris* was to be had in Dalecarlia? or mines of the very beſt kind of *gold* in Smolandia?

You will perhaps ſcarce believe me when i tell you, that there are whole mountains full of **petroleum** in Dalecarlia. Yet doubt not. This thing

thing hitherto unheard of, unseen, i myself saw with these eyes, and was surprised.

We admire the abundance of *coral* on the Indian shores, yet the port of Capellus in Gothland alone equals, nay exceeds those riches of the east. I have seen deep strata of *corals* extending many furlongs, many miles along its shores.

Botanists, who have travelled over the greatest part of the globe in search of the treasures of the vegetable kingdom, have yet left many plants for us and our posterity to discover in these our regions. For there is scarce any where a greater variety of *mosses, lichens, fuci,* and *fungi,* than with us; and the most curious botanists are now diligently employed in contemplating these minute plants.

Whoever beheld, or described our *diapensia?* who the *blasia* unless Micheli alone? These two kinds of plants grow with us, and the latter especially is found in great plenty about Fahluna. What traveller, that is not totally ignorant in botany, does not go from Paris to Fontainebleau to see those very rare *orchis's,* some of which represent helmets, others knats, others flies; all of them so exactly, so wonderfully, that there seems nothing wanting to make them the very animals themselves, but noise, and motion? Who imagined these flowers grew in our countrey, and in such plenty in Oeland, that they are to be met with in every field?

Who would ever have thought of looking in our countrey for the following exotics? The *winged pea*, the *great burnet*, the *perennial lettuce*, the *dwarf carline thistle*, the *middle fleabane*, the *black hellebore*, the Illyric *crowfoot*, much less the *riccia*, and herb *terrible*, and especially the *scorpion sena*, that most beautiful shrub, which in winter is carefully guarded against the frosts in the stoves of our botanists; yet all these have lately been observed to grow in Oeland and Gothland.

We used to purchase at a great price from forreigners the following medicinal plants, *vervain*, *moneywort*, &c. which all are natives of Sweden, and yet ten years ago nobody knew this.

What expences have we been at yearly to get the *glass-wort*, of whose ashes and salt, glass is made. The *dyers weed* and *woad* were purchased yearly at a very high price; plants that we have at last found grow every where about our provinces.

Lapland

Lapland alone furnished me some time ago with a hundred rare plants. I have gathered lately as many in the islands of the Baltic, and in Scania as many more, never before observed in Sweden. Nor can it be doubted, but that our other provinces conceal in their unfrequented corners other new plants, valuable for use or beauty, tho' hitherto overlooked, if a diligent and acute inquirer be not wanting. I will not say with the poet, "Happy the rural inhabitant," but "happy the Swedish inhabitant if he knew but what good he is possessed of [b]."

The zoologist will no where meet with a place more delightful, and more suitable to his views, than that where flocks of all kinds of birds in spring time, and summer, gather

---

[b] Our countrey has been searched by so many able botanists for plants, that what is said here cannot be applied to us. But a curious traveller might be of great service in relation to plants even here, by observing and making generally known what plants are peculiarly cultivated in some countries. Thus for instance they sow *lotus.* 13. Linn. *birdsfoot trefoil*, Ray *syn.* 334. in Herefordshire, which grows all over England on dry pastures, and is found very good for sheep, tho' every where else, as far as i know, neglected. Again they make great use of the *common vetch* in Glocestershire, chiefly for horses, feeding them with it upon the spot, and eating it up time enough for turneps the same year.

together to breed. This is the case in the woody and mountainous parts of Sweden, more than in any other spot of the earth; the Lapland *plover* called *pago*, the Norland *pied chaffinch*, the Oeland *tringa* called *akwargrim*, the Gothland *duck* called *eider*, the artic *duck* of the island Carolina called *torde*, the Ottenbyensian *cobler's awl* called *sierfloecha*, the *picus tridactylus* of the Dalecarlians are all more rare in other countries than pheasants are with us. I may venture to affirm that no countrey upon the face of the earth abounds more with birds and insects, than Sweden. Wild *reindeer, flying squirrels*, and the Norway *rat* that pours down in troops from the mountains into the plains below are unknown, and perhaps happily unknown, any where else.

Forreigners come into the Dalecarlian mountains to catch *falcons*, as is well known.

In the island Farô, situated near Gothland, *whale* and *salmon* fishery is very conveniently carried on, and no where with greater profit.

How many species of fishes furnish our tables very common in Sweden, especially of the softmouthed kind; such as the *asp*, the *wimba*, the *faren*, the *biorkna*, the *mudd*, and others,

others, unknown, undescribed, unseen, in forreign countries. Who ever dissected, examined, described those minute *red serpents* called *asps*, or *æspingar* by the southern Swedes, whose bite communicates a deadly poison?

It would be tedious were i to descend to the species of insects. The great Reaumur, who has shewn a sagacity, and accuracy, before him unknown, in examining insects, upon seeing my collection of Swedish insects owned ingenuously, that my countrey alone contained more species of those animals, than any other known in the world.

The curious diætetic, whose business it is to inquire into the various ways of living among men, will scarcely find any place, where there are so many different kinds of food, as here. Here men vary in their food, as they vary in fortune, situation, and condition. And what is very remarkable, the inhabitants of this northern world have their peculiar customs, and rules of eating in every province, and territory. In Lapland they live without corn, or wine, without salt, or any made liquor. Water, and flesh, and preparations of these are their only sustenance.

In some places the countreyman lives in his smoaky, and sooty stove on the [e] *corøgonus* when stinking; and bread made of the roots of the *calla*, or of the husks, and beards of grain pounded.

In some places they live upon stinking *herring*, and *ropy* whey called *syra*; in other places on a food called *assu*, and *artsau*, and stinking *fish*; and yet they undergo much labor. In some places their food is *turneps*, and their drink made of *juniper berries*. Some live upon *peas*, others on *buck wheat*, others grow fat upon *whale*'s flesh, to the astonishment of strangers.

In travelling thro' other countries, you will hardly ever see so many different ways of living in this respect, as in the Swedish dominions, and where consequently the diætetic philosopher may have so many opportunities of making his experiments.

The pathologist, who inquires into the causes of distempers, will not lose his time in travelling into these countries; as in every dif-

---

[e] A general name of fishes, some species of which are known in England and Wales, as the *schelley*, the *grayling*, the *gwiniad*, &c. Vid. Artedi ichthyolog.

ferent province men are subject to peculiar diseases, which arise in a great measure from the different kinds of food, that prevail among them. He will no doubt hence be empowered to assign the true causes; why the Norlander is infected with the scurvy, and why the Laplander on the contrary is free from it: why the same Laplander is subject to those terrible gripes, called by them *ullem*; why the Gothlander is chiefly afflicted with the hypochondriacal colic; why the West-Bothnians, who are more prolific than any other people in our part of the world, lose most of their children in the cradle; why so many people are liable to the epilepsy in the territory of Verns, for the cause is slight in appearance, but very singular in its nature.

Why almost all the males in Orsobæa dye of consumptions before the age of 30.

To enumerate all the things which we have particular opportunities of observing in relation to these affairs in our countrey, would require no short treatise.

I am fully persuaded that it is absolutely necessary for the young physician to travel thro' his own countrey, were it only for this reason, that relying upon his own strength he
might

might daily become more diligent, gain experience, without which there can be no skill in physic, and bring the art which he professes to some degree of perfection. For it happens amongst us, and perhaps no where so frequently, that our common people have confidence in their physicians, and run in crouds to consult any one, that is known to have taken a doctor's degree; in other countries they will scarcely trust a young physician with a favorite dog.

By following this course, and entering into practice, the young physician will perceive, whether medicines, oftentimes celebrated beyond all bounds of moderation, have that effect upon the patient, which we find mentioned in practical books. He will hear of many domestic remedies, unknown elsewhere, in use among the countrey people, that are looked upon as specifics, and preferred to the most costly prescriptions; for during the consultation, the patient may reveal the secret, if the physician is prudent, and makes use of a little art.

What are those famous exotic remedies brought from either Indies, and purchased at so great a price. v. g. *sarsaparilla*, a spe-

species of *smilax, ipecacuanha,* a species of *honey-suckle, acmella,* a species of *hemp agrimony, contrayerva* of *dorstenia,* and *simoruba* of *pistacia,* which in some diseases are reckoned specifics? what are all these, i say, but remedies approved by long use amongst the vulgar? and are not innumerable remedies used among our own countrey people of the same nature? were not all those i have enumerated found out by [d] barbarians, and when experience had shewn, that they were useful, and efficacious in many diseases, were they not thought worthy to be communicated to the rest of mankind? Let our young physician then learn,

[d] Vid. *Vires plantarum Amænitat. academ.* vol. i. p. 403. where Brunnerus is quoted for saying, that barbarians have done more towards the advancement of physic, than the learned of all ages. In the same passage the following words of Tournefort are quoted, *que tout le travail des hommes n'a encore rien produit de si assuré que deux ou trois drogues que les sauvages trouvent dans les bois.* The author subjoins to these quotations a list of twenty medicines with an &c. taken from barbarous nations, now used in our shops.

The curious reader may find in Dampier's voyages a very extraordinary instance of the skill of the savages of America in the chirurgical way. Wafer there gives an account of a cure performed upon himself by these people, and his testimony is the stronger, as he was a surgeon himself.

not

not to contemn, but accurately to remark those remedies, which are cried up amongst the common people. For he who boasts of knowing more of the virtues of simples, than what ^e taste, smell, ^f fructification, and experiments will suggest, vehemently deceives, or is deceived.

Ye who intend one of these days to cultivate your native soil with advantage, and profit, may be assured that you will find nothing in all the books of husbandry, that will be of such assistance to you in that art, as travelling thro' the different provinces of this kingdom. In some parts, and those the most barren, you

^e Vid. *Amænit. academ.* vol. 2. p. 371. in an express treatise on this subject the author quotes several eminent physicians both ancient and modern, who maintained the same opinion as to tastes. And vol. 3. p. 183. where the assistance to be had from smell is considered, and the effects of odors amply treated on.

^f Fructification. The reader perhaps may be at a loss to understand this. The meaning of it is, that plants which agree in the genus and even in the class agree also in their vertues. Thus the leaves of all the grasses are good for cattle, the lesser seeds for small birds, the greater for man, and this without exception. The stellated plants of Ray are diuretic, the rough-leaved plants of the same author are astringent and vulnerary. Plants with a pea-flower are all wholesome for cattle and man, &c. Vid. a curious treatise on this subject, in the *Academ.* vol. 1. p. 389.

will

will see very confiderable crops produced by the force of fkill, and induftry. In others, tho by nature extremely fertile, you will fcarcely fee any appearance of crops ; and the inhabitants live poorly, and in a miferable condition, merely from careleffnefs, and indolence. You may obferve how far the Cuprimontani exceed all others in the management of hay, and grafs, and the Gothlanders, in relation to cattle, and particularly fheep.

You will have an opportunity of noting the different ways in different places of ploughing, manuring, harrowing, fowing, reaping, gathering, drying, and threfhing, from whence a prudent traveller may judge which way is beft.

It would be abfurd indeed to apply to our lands forreign methods of hufbandry in every particular, v. g. forreign grafs feed would not fucceed fo well as our own. Yet i will venture to fay one could fcarcely travel a day in any of our countries without learning fomething of ufe in œconomy. Many things that will occur, may appear trifling at firft fight, which yet upon a more mature confideration, you will own may be turned to very great advantage ; fuch as the various ways of cloathing, preparing victuals,

tuals, feeding cattle, not to mention the manners, commerce and numberless other particulars.

Lastly, however necessary and incumbent upon us it may be to take a view of our countrey, it will be in vain to undergo this trouble, if we do not lay the foundation of our studies at the university, as to natural philosophy, natural, and medical history; without which preparation for travelling to advantage every thing that occurs, will appear trite, common, and not worth our attention. The traveller however, above all men, ought to keep in mind that famous principle of Descartes, viz. to doubt about every thing. He must also be very cautious not to suffer his mind, from too eager a desire of knowledge, to be overwhelmed at the beginning by the number of things to be observed g. * * * * * *
* * * * * * * * * * *
We ought to travel in the flower of our age, while the mind, and body are in vigor, while our strength is unimpaired, and alacrity at its height; before a family, houshold affairs,

_g Here follow some few lines in the original, which not understanding i have omitted.

and conjugal tyes have engrossed our affections.

When by this method you have laid the first foundation of travelling in your own countrey, you will then be qualified to go farther, and become serviceable to yourselves, and the public, by learning those things abroad, which could not be learned at home; and thus, having made a fair examination, you may be enabled to judge, whether our own customs may be improved by the help of forreign ones, and how far; and thus you will not be apt rashly to imagine, that every fashion which prevails at Paris, is fit to be introduced into our cottages; lastly, thus you will not be better acquainted with the manners and customs in France, England, Germany, and other countries, than with those of your own; i. e. you will not, as the proverb says, for want of common sense,

Invert all order, and become
Lynxes abroad, mere moles at home.

But not to abuse your patience any longer, i here break off the thread of my discourse, that what time remains may be employed by me in expressing my wishes and thanks. First, to thee, O omnipotent God, i humbly offer up my thanksgiving for the immense benefits, that have been heaped upon me thro' thy gracious protection and providence. Thou from my youth upwards hast so led me by the hand, hast so directed my footsteps, that i have grown up in the simplicity, and innocence of life, and in the most ardent pursuit after knowledge. I give thee thanks for that thou hast ever preserved me in all my journies thro' my native and forreign countries, amidst so many dangers, that surrounded me on every side. That in the rest of my life, amidst the heaviest burthens of poverty, and other inconveniences, thou wast always present to support me with thy almighty assistance. Lastly that amidst so many vicissitudes of fortune, to which i have been exposed, amongst all the goods, i say, and evils, the joyfull and gloomy, the pleasing, and disagreeable circumstances of life, thou endowedst me with an equal, constant, manly, and superior spirit on every occasion.

To our most august, and potent prince Frederick the first, as becomes a dutifull, and obedient subject, i give most humble thanks for his favorable kindness in bestowing upon me this honorable post. May the almighty grant, that his majesty, and his most serene consort, those shining stars of the north, may long, very long illuminate, and adorn this region with the brightness of their rays.

To thee, most mighty count Gyllenbourg, illustrious chancellor of this university, to thee, though absent, i return the most sincere, and humble thanks for the great, and even endless benefits bestowed upon me; amongst which, exceeding all number, this must not be reckoned the least, that, when i was called hither by this academy, you recommended me in the most indulgent manner to our great monarch. It shall be my constant care that you may never repent of this favor, and by reverence, respect, and duty, to testify my gratitude to my latest breath.

To the most reverend the archbishop, to the vice-chancellor, to the magnificent rector, and to you illustrious and celebrated professors, i return also most grateful acknowledgments, who honored me by your unanimous votes, and assisted in bringing me to this chair. As this your benevolence laid me under the greatest obligation to you, to employ every office of regard and friendship towards you, so by the grace of God i shall omit no opportunity of shewing i am not unworthy of your favour.

Whilst i am thus employed in testifying the feelings of a grateful mind, i ought not to forget your name, most illustrious Roberg, my predecessor highly worthy of the utmost veneration. As i am one of those who have had the happiness of being educated in your school, i should be the most ungratefull of men, if i were ever to suffer the remembrance of such a benefit to slip out of my mind.

It has been your lot, venerable sir, to survive all your brethren; and you may justly boast, what every physician now in Sweden will gratefully confess, that to you, as to their faithful instructor, they owe the beginning, the increase, and the finishing of their art. Nay not only the faculty at Upsal, but the whole circle here present ought to salute, and reverence you as superior to them all in age.

Suffer then at last your disciple to ease you of that burthen, which for forty years, and more, you have sustained, with honor; that now, time having laid his heavy hand upon you, you may enjoy that rest, which his sacred majesty has kindly granted to your wishes. My sincere prayers are not wanting to the almighty, that he may grant you a chearfull, and vigorous old age, and that every

every thing may succeed to the utmost of your desires.

Nor is it fit, ye florishing and chosen youth, that on this solemn occasion i should pass you over in silence. I have been long sensible of your regard for me, by many and undoubted proofs; i have been long sensible, i say, and i gratefully acknowledge it. Many of you desired, ardently desired to have me in this station, though perhaps never seen by you before. For this alone, i know, i am called hither, that i may be useful to you. On you therefore my fortune turns. My industry, my studies, my labors, my watchings, i willingly, and chearfully consecrate to your service, and by the assistance of God, i will exert the utmost of my power to satisfy your expectations, that you may not be disappointed in the hopes you have conceived of me.

THE
OECONOMY
OF
NATURE.

THE OECONOMY OF NATURE.

BY

ISAAC J. BIBERG.

Upsal, 1749. March 4.

Amænitat. Academ. vol. ii.

Æternæ sunt vices rerum. Sen. nat. 3. 1.

§. 1.

BY the Œconomy of Nature we understand the all-wise disposition of the Creator in relation to natural things, by which they are fitted to produce general ends, and reciprocal uses.

All things contained in the compass of the universe declare, as it were, with one accord the infinite wisdom of the Creator. For whatever strikes our senses, whatever is the object of our thoughts, are so contrived, that they concur to make manifest the divine glory, i. e. the ultimate end which God proposed in all his works. Whoever duly turns his attention to the things on this our terraqueous globe, must necessarily confess, that they are so connected, so chained together, that they all aim at the same end, and to this end a vast number of intermediate ends are subservient. But as the intent of this treatise will not suffer me to consider them all, i shall at present only take notice of such as relate to the preservation of natural things. In order therefore to perpetuate the established course of nature in a continued series, the divine wisdom has thought fit, that all living creatures should constantly be employed in producing individuals; that all natural things should contribute and lend a helping hand to preserve every species; and lastly, that the death and destruction of one thing should always be subservient to the restitution of another. It seems to me that a greater subject than this cannot be found, nor one on which laborious men may more worthily employ their industry, or men of genius their penetration.

I am

OF NATURE.

I am very sensible, being conscious of my own weakness, how vast and difficult a subject it is, and how unable i am to treat it as it deserves; a subject which would be too great a task for the ability of the most experienced and sagacious men, and which properly performed would furnish materials for large volumes. My design therefore is only to give a summary view of it, and to set forth to the learned world, as far as i am able, whatever curious, worthy to be known, and not obvious to every observer occurs in the triple kingdom of nature. Thus if what the industry of others shall in future times discover in this way be added to these observations, it is to be hoped that a common stock may thence grow, and come to be of some importance. But before i examine these three kingdoms of nature, it will not, i think, be amiss to say something concerning the earth in general, and its changes.

§. 2.

The world, or the terraqueous globe, which we inhabit is every where surrounded with elements, and contains in its superficies the three kingdoms of nature, as they are called; the *fossil*, *fil,* which constitutes the crust of the earth, the *vegetable,* which adorns the face of it, and draws the greatest part of its nourishment from the *fossil* kingdom, and the *animal,* which is sustained by the *vegetable* kingdom. Thus these three kingdoms cover, adorn and vary the superficies of our earth. It is not my design to make any inquiry concerning the center of the terraqueous globe. He, who likes hypotheses, may consult Descartes, Helmont, Kircher, and others. My business is to consider the external parts of it only, and whatever is obvious to the eye.

As to the *strata* of the earth and mountains, as far as we have hitherto been able to discover, the upper parts consist of *rag-stone,* the next of *slate,* the third of *marble* filled with petrifactions, the fourth again with *slate,* and lastly the lowest of *free-stone.* The habitable part of the earth, though it is scooped into various inequalities, yet is every where high in comparison with the water, and the farther it is from the sea it is generally higher. Thus the waters in the lower places are not at rest, unless some obstacle confines them, and by that means form lakes and marshes.

The sea surrounds the continent, and takes up the greatest part of the earth's superficies, as geography informs us. Nay, that it once spread over much the greatest part, we may be convinced by its yearly decrease, by the rubbish left by the tides, by *shells*, *strata*, and other circumstances.

The sea-shores are usually full of dead testaceous animals, wrack, and such like bodies, which are yearly thrown out of the sea. They are also covered with sand of various kinds, stones, and heaps of other things not very common. It happens moreover, that while the more rapid rivers rush through narrow vallies, they wear away the sides, and thus the friable, and soft earth falls in, and its ruins are carried to distant, and winding shores; whence it is certain, that the continent gains no small increase, as the sea subsides.

The clouds collected from exhalations, chiefly from the sea, but likewise from other waters, and moist grounds, and condensed in the lower regions of the atmosphere, supply the earth with rain; but since they are attracted by the mountainous parts of the earth, it necessarily follows, that those parts must have, as is fit, a larger share of water than the rest. Springs, which generally rush out at the foot of mountains, take their rise from

this

this very rain water, and vapours condenfed, that trickle through the holes, and interftices of loofe bodies, and are received into caverns.

These afford a pure water purged by ftraining, which rarely dry up in fummer, or freeze in winter, fo that animals never want a wholefome and refrefhing liquor.

The chief fources of rivers are fountains, and rills growing by gradual fupplies into ftill larger and larger ftreams, till at laft, after the conflux of a vaft number of them, they find no ftop, but falling into the fea with leffened rapidity, they there depofit the united ftores they have gathered, along with foreign matter, and fuch earthy fubftances, as they tore off in their way. Thus the water returns in a circle, whence it firft drew its origin, that it may act over the fame fcene again.

Marfhes arifing from water retained in low grounds are filled with moffy tumps, which are brought down by the water from the higher parts, or are produced by putrified plants.

We often fee new meadows arife from marfhes dryed up. This happens fooner when the g *fphagnum* F. S. 864. * has laid a founda-

  g A kind of mofs.
  * This refers to the firft Edition of the *Flora Suecica*. It is 953 in the fecond Edition.

tion;

tion; for this in procefs of time changes into a very porous mould, till almoſt the whole marſh is filled with it. After that the *ruſh* ſtrikes root, and along with the *cotton graſſes* conſtitutes a turf, raiſed in ſuch a manner, that the roots get continually higher, and thus lay a more firm foundation for other plants, till the whole marſh is changed into a fine and delightfull meadow; eſpecially if the water happens to work itſelf a new paſſage.

Hillocks, that abound in low grounds, occaſion the earth to encreaſe yearly, more than the countreyman would wiſh, and ſeem to do hurt: but in this the great induſtry of nature deſerves to be taken notice of. For by this means the barren ſpots become ſooner rich meadow, and paſture land. Theſe hillocks are formed by the ant, by ſtones, and roots, and the trampling of cattle; but the principal cauſe is the force of the winter cold, which in the ſpring raiſes the roots of plants ſo high above the ground, that being expoſed to the air they grow, and periſh; after which the *golden maidenhairs* fill the vacant places.

Mountains, hills, vallies, and all the inequalities of the earth, though ſome think they take away much from its beauty, are ſo far from producing such an effect, that on the contrary they give a more pleasing aspect, as well as great advantages. For thus the terrestrial superficies is larger; different kinds of plants thrive better, and are more easily watered, and the rain-waters run in continual streams into the sea, not to mention many other uses in relation to winds, heat and cold. Alps are the highest mountains, that reach to the second region of the air, where trees cannot grow erect. The higher these Alps are, the colder they are *cæteris paribus*. Hence the Alps in Sweden, Siberia, Swifferland, Peru, Brasil, Armenia, Asia, Africa, are perpetually covered with snow; which becomes almost as hard as ice. But, if by chance the summer heats be greater than ordinary, some part of these stores melts, and runs through rivers into the lower regions, which by this means are much refreshed.

It is scarcely to be doubted, but that the rocks and stones dispersed over the globe were formed originally in and from the earth; but when torrents of rain have softened, as they easily do, the soluble earth, and carried it down into the lower parts, we imagine it happens that these solid, and heavy bodies, being laid

OF NATURE. 47

laid bare, stick out above the surface. We might also take notice of the wonderfull effects of the tide, such as we see happen from time to time on the sea-shore, which being daily and nightly assaulted with repeated blows, at length gives way and breaks off. Hence we see in most places the rubbish of the sea, and shores.

The winter by its frost prepares the earth, and mould, which thence are broken into very minute particles, and thus, being put into a mouldering state, become more fit for the nourishment of plants; nay by its snow it covers the seeds, and roots of plants, and thus by cold defends them from the force of cold. I must add also that the piercing frost of the winter purifies the atmosphere, and putrid waters, and makes them more wholesome for animals.

The perpetual succession of heat and cold with us renders the summers more pleasing; and though the winter deprives us of many plants and animals, yet the perpetual summer within the tropics is not much more agreeable, as it often destroys men, and other animals by its immoderate heat; though it must be confessed

fessed that those regions abound with most exquisite fruits. Our winters, though very troublesome to a great part of the globe, on account of their vehement and intense cold, yet are less hurtfull to the inhabitants of the northern parts, as experience testifies. Hence it happens, that we may live very conveniently on every part of the earth, as every different countrey has different advantages from nature.

The seasons, like every thing else, have their vicissitudes, their beginnings, their progress, and their end.

The age of man begins from the cradle, pleasing childhood succeeds, then active youth, afterwards manhood firm, severe and intent upon self-preservation, lastly old age creeps on, debilitates, and at length totally destroys our tottering bodies.

The seasons of the year proceed in the same way. Spring, the jovial, playfull infancy of all living creatures, represents childhood and youth; for then plants spread forth their luxuriant flowers, fishes exult, birds sing, every part of nature is intent upon generation. The summer, like middle age, exhibits plants, and trees every where cloathed with green; it
gives

gives vigor to animals, and plumps them up, fruits then ripen, meadows look cheerfull, every thing is full of life. On the contrary autumn is gloomy, for then the leaves of trees begin to fall, plants to wither, infects to grow torpid, and many animals to retire to their winter quarters. The day proceeds with juft fuch fteps, as the year. The morning makes every thing alert, and fit for bufinefs; the fun pours forth his ruddy rays, the flowers which had, as it were, flept all night, awake and expand themfelves again; the birds with their fonorous voices, and various notes, make the woods ring, meet together in flocks, and facrifice to Venus. Noon tempts animals into the fields, and paftures; the heat puts them upon indulging their eafe, and even neceflity obliges them to it. Evening follows, and makes every thing more fluggifh; flowers fhut up,[h] and animals retire to their lurking

[h] Of fuch flowers as fleep by night fome account is given by Linnæus in Philof. Botan. p. 88. where the curious may alfo find, p. 274, a lift of plants one or other of which fhut their flowers at every hour of the day without regard to the weather. One plant is fo remarkable for this property, that it is generally known in our countrey by the name of go-to-bed at-noon. Its botanical name is tragopogon or goat's-beard. See a Differtation in the Amæn. acad. vol. 4. where this fubject is treated at large.

E   places.

places. Thus the spring, the morning, and youth are proper for generation; the summer, noon, and manhood are proper for preservation; and autumn, evening, and old age are not unfitly likened to destruction.

§. 3.

The fossil kingdom.

Propagation.

It is agreed on all hands, that stones are not organical bodies, like plants, and animals; and therefore it is as clear that they are not produced from an egg, like the tribes of the other kingdoms. Hence the variety of fossils is proportionate to the different combinations of coalescent particles, and hence the species in the fossil kingdom are not so distinct, as in the other two. Hence also the laws of generation in relation to fossils have been in all ages extremely difficult to explain; and lastly hence have arisen so many different opinions about them, that it would be endless to enumerate them all. We therefore for the present will content ourselves with giving a very few observations on this subject.

That

That clay is the sediment of the sea is sufficiently proved by observation, for which reason it is generally found in great plenty along the coasts.

The journals of seamen clearly evince, that a very minute sand covers the bottom of the sea, nor can it be doubted, but that it is daily crystallised out of the water.

It is now acknowledged by all, that testaceous bodies and petrifactions resembling plants were once real animals or vegetables;[1] and it seems likely that shells being of a calcareous nature have changed the adjacent clay, sand, or mould into the same kind of substance. Hence we may be certain, that marble may be generated from petrifactions, and therefore it is frequently seen full of them.

Rag-stone the most common matter of our rocks appears to be formed from a sandy kind of clay, but this happens more frequently, where the earth is impregnated with iron.

Freestone is the product of sand, and the deeper the bed, where it is found, the more compact it becomes; and the more dense the

---

[1] I have taken the liberty not to follow the original text in this place. The learned will see the reason at first sight.

sand, the more easily it concretes. But if an *alcaline* clay chances to be mixed with the sand, the *freestone* is generated more readily, as in the *freestone* called *cos friatilis, particulis argillo-glarensis*, S. N. 1. 1.

The *flint*, S. N. 3. 1. is almost the only kind of *stone*, certainly the most common in *chalky* mountains. It seems therefore to be produced from *chalk*. Whether it can be reduced again to *chalk*, i leave to others to inquire.

Stalactites, S. N. 33. 1. or *drop-stone* is composed of calcareous particles, adhering to a dry and generally a vegetable body.

The incrustations S. N. 32. 5, 6, 7, 8. are often generated, where a vitriolic water connects claiey and earthy particles together.

Slate by the vegetables, that are often inclosed in it, seems to take its origin from a marshy mould.

Metals vary according to the nature of the matrix, to which they adhere, e. g. the *pyrites cupri Fahlunensis* contains frequently *sulphur, arsenic, iron, copper*, a little *gold, vitriol, alum*, sometimes *lead ore, silver* and *zinck*. Thus *gold, copper, iron, zinck, arsenic, pyrites, vitriol*, come out of the same vein. That very rich *iron* ore at Normark in Vermelandia, where

it

OF NATURE.

it was cut transversly by a vein of clay, was changed into a pure *silver*. The number therefore of species, and varieties of *fossils*, each serving for different purposes, according to their different natures, will be in proportion, as the different kinds of earths and *stones* are variously combined.

§. 4.

Preservation.

As *fossils* are destitute of life, and organisation, are hard, and not obnoxious to putrefaction; so they last longer, than any other kind of bodies. How far the air contributes to this duration it is easy to perceive, since air hardens many *stones* upon the superficies of the earth, and makes them more solid, compact, and able to resist the injuries of time. Thus it is known from vulgar observation that *lime*, that has been long exposed to the air, becomes hardened. The *chalky marl*, which they use in Flanders for building houses, as long as it continues in the quarry, is friable; but when dug up and exposed to the air, it grows gradually harder. In the same way our old walls, and towers gain a firmness in process of time, and therefore it is a vulgar miſtake, that our anceſtors excelled the modern architects in the art of building as to this point [k].

However ignorant we may be of the cauſe, why large rocks are every where to be ſeen ſplit, whence vaſt fragments are frequently torn off; yet this we may obſerve, that fiſſures are cloſed up by water, that gets between them, and is detained there; and are conſolidated by *cryſtal* and *ſpar*. Hence we ſcarcely ever find *cryſtal*, but in thoſe *ſtones*, which have for ſome time in its chinks water loaded with ſtony particles. In the ſame manner *cryſtals* fill the cavities in mines, and concrete into *quartz* or a debaſed *cryſtal*.

It is manifeſt that *ſtones* are not only generated, augmented, and changed perpetually

---

[k] Too great ſtreſs ought not, I think, to be laid on this obſervation of our author, though it may be in part true; for without ſuppoſing that our anceſtors had more ſkill in building, we may ſuppoſe, what was likely to be the caſe, that they uſed more care in the choice of their materials, and had them wrought up with more labor; which muſt add conſiderably to the firmneſs of the cement. Where theſe circumſtances have happened to be wanting, time alone has not been able to produce the ſame effect. I have ſeen a houſe about fourſcore years old, where one might rub out the mortar from between the bricks without ſcarcely uſing any force.

OF NATURE.

from incrustations brought upon moss, but are also increased by *crystal* and *spar*. Not to mention that the adjacent earth, especially if it be impregnated with iron particles, is commonly changed into a solid *stone*.

It is said, that the *marble quarries* in Italy, from whence fragments are cut, grow up again. *Ores* grow by little and little, whenever the mineral particles, conveyed by the means of water through the clefts of mountains, are retained there; so that adhering to the homogeneous matter a long while, at last they take its nature, and are changed into a similar substance.

§. 5.

Destruction.

Fossils, although they are the hardest of bodies, yet are found subject to the laws of destruction, as well as all other created substances. For they are dissolved in various ways by the elements exerting their force upon them, as by water, air and the solar rays, as also by the rapidity of rivers, violence of cataracts, and eddies which continually beat upon and at last reduce to powder the hardest *rocks*. The agitations

tations of the sea, and lakes, and the vehemence of the waves, excited by turbulent winds pulverise *stones*, as evidently appears by their roundness along the shore. Nay as the poet says,

The hardest stone insensibly gives way

To the soft drops that frequent on it play.
So that we ought not to wonder, that these very hard bodies moulder away into powder, and are obnoxious like others to the consuming tooth of time.

Sand is formed of *freestone*, which is destroyed partly by frost, making it friable, partly by the agitation of water, and waves; which easily wear away, dissolve, and reduce into minute particles, what the frost had made friable.

Chalk is formed of rough *marble*, which the air, the sun, and the winds have dissolved, as appears by Iter. Goth. 170.

The *slate* earth or *humus schisti* Syf. Nat. 511. owes its origin to *slate*, dissolved by the air, rain and snow.

Ochre is formed of metals dissolved, whose *faces* present the very same colours, which we always find the *ore* tinged with, when exposed to the air. *Vitriol* in the same
man-

manner mixes with water from *ores* destroyed.

The *muria saxatilis* Syf. Nat. 14. 6. a kind of talky ftone yielding falt in the parts that are turned to the fun, is diffolved into fand, which falls by little and little upon the earth, till the whole is confumed; not to mention other kinds of *foffils*. Laftly from thefe there arife new *foffils*, as we mentioned before, fo that the deftruction of one thing ferves for the generation of another.

Teftaceous worms ought not to be paffed over on this occafion, for they eat away the hardeft *rocks*. That fpecies of *fhell fifh* called the *razor fhell* bores thro' ftones in Italy, and hides itfelf within them; fo that the people who eat them are obliged to break the ftones, before they can come at them. The *cochlea* F. S. 1299. a kind of *fnail* that lives on craggy *rocks*, eats, and bores through the chalky hills, as worms do through wood. This is made evident by the obfervations of the celebrated de Geer.

§. 6.

§. 6.

The Vegetable Kingdom.

Propagation.

Anatomy abundantly proves, that all *plants* are *organic*, and living bodies; and that all *organic* bodies are propagated from an egg has been sufficiently demonstrated by the industry of the moderns. We therefore the rather, according to the opinion of the skilfull, reject the æquivocal generation of *plants*; and the more so, as it is certain that every living thing is produced from an egg. Now the seeds of *vegetables* are called eggs; these are different in every different *plant*, that the means being the same, each may multiply its species, and produce an offspring like its parent. We do not deny, that very many *plants* push forth from their roots fresh offsets for two or more years. Nay not a few *plants* may be propagated by branches, buds, suckers and leaves fixed in the ground, as likewise many trees. Hence their stems being divided into branches, may be looked on as roots above ground; for in the same way the roots creep under ground;

and

OF NATURE.

and divide into branches. And there is the more reason for thinking so, because we know that a tree will grow in an inverted situation, viz. the roots being placed upwards, and the head downwards, and buried in the ground; for then the branches will become roots, and the roots will produce leaves and flowers. The *lime-tree* will serve for an example, on which gardeners have chiefly made the experiment. Yet this by no means overturns the doctrine, that all *vegetables* are propagated by seeds; since it is clear that in each of the foregoing instances nothing vegetates but what was part of a plant, formerly produced from seed, so that, accurately speaking, without seed no new *plant* is produced.

Thus again plants produce seeds, but they are entirely unfit for propagation, unless fœcundation precedes, which is performed by an intercourse between different sexes, as experience testifies. *Plants* therefore must be provided with *organs* of generation; in which respect they hold an analogy with *animals*. Since in every *plant* the flower always precedes the fruit, and the fœcundated seeds visibly arise from the fruit; it is evident that the *organs* of generation are contained in the flower,

which

which *organs* are called *antheræ*, and *stigmata*, and that the impregnation is accomplished within the flower. This impregnation is performed by means of the dust of the *antheræ* falling upon the moist *stigmata*, where the dust adheres, is burst, and sends forth a very subtle matter, which is absorbed by the *style*, and is conveyed down to the rudiment of the seed, and thus renders it fertile. When this operation is over, the *organs* of generation wither and fall, nay a change in the whole flower ensues. We must however observe, that in the *vegetable kingdom* one, and the same flower does not always contain the *organs* of generation of both sexes, but oftentimes the male *organs* are on one *plant*, and the female on another. But that the business of impregnation may go on successfully, and that no plant may be deprived of the necessary dust, the whole most elegant *apparatus* of the *antheræ* and *stigmata* in every flower is contrived with wonderful wisdom.

For in most flowers the *stamina* surround the *pistills*, and are about the same height; but there are many *plants*, in which the *pistill* is longer than the *stamina*, and in these it is wonderful to observe, that the Creator

has

OF NATURE.

has made the flowers recline, in order that the dust may more easily fall into the *stigma*, e. g. in the *campanula*, *cowslip*[1], &c. But when the fœcundation is compleated the flowers rise again, that the seeds may not fall out before they are ripe, at which time they are dispersed by the winds. In other flowers on the contrary the *pistill* is shorter, and there the flowers preserve an erect situation, nay when the flowering comes on they become erect; tho' before they were drooping, or immersed under water. Lastly, whenever the male flowers are placed below the female ones, the leaves are exceedingly small, and narrow, that they may not hinder the dust from flying upwards, like smoak; as we see in the *pine*, *fir*, *yew*, *sea-grape*, *juniper*, *cypress*, &c. and when in one and the same species one plant is male and the other female, and consequently may be far from one another, there the dust, without which there is no impregnation, is carried in abundance by the help

[1] This curious phænomenon did not escape the poetical eye of Milton, who was so very much struck with the beauty of it, that he thought it worth describing in the following enlivened imagery,

With cowslips *wan that hang the pensive head.*

of the wind from the male to the female; as in the whole *dioicous* [m] clafs. Again a more difficult impregnation is compenfated by the longævity of the individuals, and the continuation of life by buds, fuckers and roots, fo that we may obferve every thing moft wifely difpofed in this affair. Moreover we cannot without admiration obferve that moft flowers expand themfelves when the fun fhines forth, whereas when clouds, rain, or the evening comes on, they clofe up, left the genital duft fhould be coagulated, or rendered ufelefs, fo that it cannot be conveyed to the *ftigmata*. But what is ftill more remarkable and wonderfull! when the fœcundation is over, the flowers neither upon fhowers, nor evening coming on clofe themfelves up. Hence when rain falls in the flowering time, the hufbandman and gardener foretell a fcarcity of fruits. I could and would illuftrate all this by many remarkable inftances, if the fame fubject had not lately been explained, in this very place [n] in a manner equal to its

[m] i. e. where one plant bears male flowers, and the other female ones.

[n] I fuppofe the author here alludes to a treatife publifhed in Amæn academ. vol. 1. entitled, *Sponfalia plantarum*, in which are contained fo many proofs of the reality of the different fexes of plants, that to me there feems to remain no room for doubt.

importance. I cannot help remarking one particular more, viz. that the organs of generation, which in the animal kingdom are by nature generally removed from sight, in the vegetable are exposed to the eyes of all, and that when their nuptials are celebrated, it is wonderfull what delight they afford to the spectator by their most beautiful colours and delicious odors. At this time bees, flies, and other insects suck honey out of their nectaries, not to mention the humming bird; and that from their effete dust the bees gather wax.

§. 7.

As to the dissemination of seeds, after they come to maturity, it being absolutely neccisary; since without it no crop could follow; the Author of nature has wisely provided for this affair in numberless ways. The stalks and stems favor this purpose, for these raise the fruit above the ground, that the winds, shaking them to and fro, may disperse far off the ripe seeds. Most of the [n] *pericarps* are shut at

[n] Whatever surrounds the seeds is called by botanical writers a *pericarpium*, and as we want an English word to express this, i have taken the liberty to call it a pericarpy.

top,

top, that the seeds may not fall, before they are shook out by stormy winds. Wings are given to many *seeds*, by the help of which they fly far from the mother plant, and oftentimes spread over a whole countrey. These wings consist either of a down, as in most of the composite flowered plants, or of a membrane, as in the *birch*, *alder*, *ash*, &c. Hence *woods*, which happen to be consumed by fire, or any other accident, will soon be restored again by new plants, disseminated by this means. Many kinds of fruits are endued with a remarkable elasticity, by the force of which, the ripe *pericarps* throw the *seeds* to a great distance, as the *wood-sorrel*, the *spurge*, the *phyllanthus*, the *dittany*. Other *seeds* or *pericarps* are rough, or provided with hooks; so that they are apt to stick to animals, that pass by them, and by this means are carried to their holes where they are both sown, and manured by nature's wonderfull care; and therefore the plants of these *seeds* grow, where others will not, as *hounds-tongue*, *agrimony*, &c.

Berries and other *pericarps*, are by nature allotted for aliment to animals, but with this condition, that while they eat the pulp they shall sow their *seeds*; for when they feed upon it they either disperse them at the same time, or, if they swallow them, they are returned with interest; for they always come out unhurt. It is not therefore surprising, that if a field be manured with recent mud or dung not quite rotten, various other plants, injurious to the farmer, should come up along with the grain, that is sowed. Many have believed that *barley*, or *rye* has been changed into *oats*, altho' all such kinds of metamorphoses are repugnant to the laws of generation, not considering that there is another cause of this phænomenon, viz. that the ground perhaps has been manured with horse-dung, in which the *seeds* of *oats*, coming entire from the horse, lye hid and produce that grain. The *misletoe* always grows upon other trees, because the thrush that eats the seeds of it, casts them forth with its dung, and as bird-catchers make their bird-lime of this same plant, and daub the branches of trees with it, in order to catch the thrush, the proverb hence took its rise;

The thrush, when he befouls the bough,
Sows for himself the seeds of woe.

It is not to be doubted, but that the greatest part of the *junipers* also, that fill our woods, are sown by thrushes, and other birds in the same manner; as the berries, being heavy, cannot be dispersed far by the winds. The cross-bill that lives on the fir-cones, and the hawfinch that feeds on the pine-cones, at the same time sow many of their *seeds*, especially when they carry the *cone* to a stone, or trunk of a tree, that they more easily strip it of its scales. Swine likewise, by turning up the earth, and moles by throwing up hillocks, prepare the ground for seeds in the same manner, as the ploughman does.

I pass over many other things, which might be mentioned concerning the sea, lakes, and rivers, by the help of which oftentimes *seeds* are conveyed unhurt to distant countries; nor need I mention in what a variety of other ways nature provides for the dissemination of plants, as this subject has been treated on at large in our illustrious president's oration concerning the augmentation of the habitable earth. [p]

§. 8.

---

[p] As there is something very ingenious, and quite new in the treatise here referred to, i will for the sake of those, who cannot read the original, give a short abstract of it. His design is to shew that there was only one pair of all living things, created at the beginning. According to the
account

§. 8.

Preservation.

The great Author and Parent of all things, decreed, that the whole earth should be covered with plants, and that no place should be void,

account of Moses, says the author, we are sure, that this was the case in the human species; and by the same account we are informed that this first pair was placed in Eden, and that Adam gave names to all the animals. In order therefore that Adam might be enabled to do this, it was necessary that all the species of animals should be in paradise; which could not happen unless also the species of vegetables had been there likewise. This he proves from the nature of their food, particularly in relation to insects, most of which live upon one plant only. Now had the world been formed in its present state, it could not have happened that all the species of animals should have been there. They must have been dispersed over all the globe, as we find they are at present, which he thinks improbable for other reasons which I shall pass over for the sake of brevity. To solve all the phænomena then he lays down a principle, that at the beginning all the earth was covered with sea, unless one island large enough to contain all animals and vegetables. This principle he endeavors to establish by several phænomena which make it probable, that the earth has been and is still gaining upon the sea, and does not forget to mention *fossil shells*, and *plants* every where found, which he says cannot be accounted for by the deluge. He then undertakes to shew

how

void, none barren. But since all countries have not the same changes of seasons, and every soil is not equally fit for every plant, He therefore, that no place should be without

how all vegetables and animals might in this island have a soil and climate proper for each, only by supposing it to be placed under the æquator, and crowned with a very high mountain. For it is well known that the same *plants* are found on the Swiss, the Pyrenean, the Scotch alps, on Olympus, Lebanon, Ida, as on the Lapland and Greenland alps. And Tournefort found at the bottom of mount Ararat the common plants of Armenia, a little way up those of Italy, higher those which grow about Paris, afterwards the Swedish plants, and lastly on the top the Lapland *alpine plants*; and i myself, adds the author, from the plants growing on the Dalecarlian alps could collect how much lower they were than the alps of Lapland. He then proceeds to shew how from one plant of each species the immense number of individuals now existing might arise. He gives some instances of the surprising fertility of certain plants, v. g. the elecampane, one plant of which produced 3000 seeds, of spelt 2000, of the sunflower 4000, of the poppy 3200, of tobacco 40320. But supposing any annual plant producing yearly only two seeds, even of this after 20 years there would be 1,048,576 individuals. For they would increase yearly in a duple proportion, viz. 2, 4, 8, 16, 32, &c. He then gives some instances of plants brought from America, that are now become common over many parts of Europe. Lastly he enters upon the subject for which he is quoted in the text, where the detail he gives of the several methods which nature has taken to propagate vegetables is extremely curious, but too long to insert in this place.

some,

OF NATURE.

some, gave to every one of them such a nature, as might be chiefly adapted to the climate; so that some of them can bear an intense cold, others an equal degree of heat; some delight in dry ground, others in moist, &c. Hence the same plants grow only where there are the same seasons of the year, and the same soil.

The *alpine* plants live only in high, and cold situations, and therefore often on the *alps* of Armenia, Switzerland, the Pyreneans, &c. whose tops are equally covered with eternal snows, as those of the Lapland *alps*, plants of the same kind are found, and it would be in vain to seek for them any where else. It is remarkable in relation to the *alpine* plants, that they blow, and ripen their *seeds* very early, otherwise the winter would steal upon them on a sudden, and destroy them.

Our northern plants, altho' they are extremely rare every where else, yet are found in Sibiria, and about Hudson's bay, as the *arbutus*, *Flor*. 339. *bramble*, 412. *wintergreen*, &c.

Plants impatient of cold live within the torrid zones; hence both the Indies tho' at such a distance from one another have plants in common.

mon. The Cape of Good Hope, i know not from what cause, produces plants peculiar to itself, as all the *mesembryanthema*, and almost all the species of *aloes*. *Grasses*, the most common of all plants, can bear almost any temperature of air, in which the good providence of the Creator particularly appears; for all over the globe they above all plants are necessary for the nourishment of cattle, and the same thing is seen in relation to our most common grains.

Thus neither the scorching sun, nor the pinching cold hinders any countrey from having its vegetables. Nor is there any soil, which does not bring forth many kinds of plants; the *pond-weeds*, the *water-lily*, *lobelia* inhabit the waters. The *fluviales, fuci, conservæ* cover the bottoms of rivers, and sea. The *sphagna* [q] fill the marshes. The *brya* [r] cloath the plains. The dryest woods and places scarce ever illuminated by the rays of the sun are adorned with the *hypna*. Nay stones and the trunks of trees are not excepted, for these are covered with various kinds of *liverwort*.

The desart, and most sandy places have their peculiar trees, and plants; and as rivers or

[q] Kind of moss.   [r] Kind of moss.

brooks

brooks are very seldom found there, we cannot without wonder observe that many of them distill water, and by that means afford the greatest comfort both to man, and beasts that travel there. Thus the * *tillandsia*, which is a *parasitical plant*, and grows on the tops of trees in the deserts of America, has its leaves turned at the base into the shape of a pitcher, with the extremity expanded ; in these the rain is collected, and preserved for thirsty men, birds, and beasts.

The *water-tree* in Ceylon produces cylindrical bladders, covered with a lid ; into these is secreted a most pure, and refreshing water, that tastes like nectar to men, and other animals. There is a kind of *cuckow-pint* in New France, that if you break a branch of it, will afford you a pint of excellent water. How wise, how beautiful is the agreement between the plants of every countrey, and its inhabitants, and other circumstances !

* A kind of *misletoe*.

§. 9.

Plants oftentimes by their very structure contribute remarkably both to their own preservation, and that of others. But the wisdom of the Creator appears no where more than in the manner of growth of *trees*. For as their roots descend deeper, than those of other *plants*, provision is thereby made, that they shall not rob them too much of nourishment; and what is still more, a stem not above a span in diameter often shoots up its branches very high; these bear perhaps many thousand buds, each of which is a *plant* with its *leaves, flowers* and *stipulæ*. Now if all these grew upon the plain, they would take up a thousand times as much space, as the *tree* does, and in this case there would scarcely be room in all the earth for so many *plants*, as at present the *trees* alone afford. Besides *plants* that shoot up in this way are more easily preserved from cattle by a natural defence, and farther their leaves falling in autumn cover the *plants* growing about against the rigor of the winter, and in the summer they afford a pleasing shade, not only to animals, but to *plants*, against the intense heat of the sun. We may add that

trees

trees like all other *vegetables* imbibe the water from the earth, which water does not circulate again to the root, as the ancients imagined [t]; but being dispersed, like small rain, by the transpiration of the leaves, moistens the *plants* that grow about. Again, many *trees* bear fleshy fruits of the *berry* or *apple* kind, which, being secure from the attack of cattle, grow ripe for the use of man and other animals, while their seeds are dispersed up and down after digestion. Lastly the particular structure of *trees* contributes very much to the propagation of insects; for these chiefly lay their eggs upon their leaves, where they are secure from the reach of cattle.

Ever-green trees, and *shrubs* with us are chiefly found in the most barren woods, that they may be a shelter to animals in the winter. They lose their leaves every third year, as their seeds are sufficiently guarded by the *mosses*, and do not want any other covering. The *palms* in the hot countries perpetually keep their leaves, for there the seeds stand in no need of any shelter whatever.

---

[t] See *Vegetable Statics* by that great philosopher, Dr. Hales, where this subject is treated in a masterly way.

Many *plants*, and *shrubs* are armed with thorns, e. g. the *buck-thorn, sloe, carduus, cotton-thistle*, &c. that they may keep off the animals, which otherwise would destroy their fruit. These at the same time cover many other *plants*, especially of the annual kind, under their branches [u]. So that while the adjacent grounds are robbed of all *plants* by the voracity of animals, some may be preserved, to ripen flowers and fruit, and stock the parts about with seeds, which otherwise would be quite extirpated.

All *herbs* cover the ground with their leaves, and by their shade hinder it from being totally deprived of that moisture, which is necessary to their nourishment. They are moreover an ornament to the earth, especially as leaves have a more agreeable verdure on the upper, than the under side.

The *mosses*, which adorn the most barren places, at the same time preserve the lesser plants, when they begin to shoot, from cold and drought. As we find by experience in our

[u] This observation may be extended farther; for it is constantly seen upon commons, where *furze* grows, that wherever there was a bush left untouched for years by the commoners, some *tree* has sprung up, being secured by the prickles of that *shrub* from the bite of the cattle.

gardens, that *plants* are preserved in the same way. They also hinder the fermenting earth from forcing the roots of *plants* upwards in the spring; as we see happen annually to trunks of trees, and other things put into the ground. Hence very few *mosses* grow in the warmer climates, as not being so necessary to that end in those places.

The English *sea mat-weed* or *marran* will bear no soil but pure sand, which nature has allotted to it. Sand the produce of the sea, is blown by winds oftentimes to very remote parts, and deluges, as it were, woods and fields. But where this grass grows, it frequently fixes the sand, gathers it into hillocks, and thrives so much, that by means of this alone, at last an entire hill of sand is raised. Thus the sand is kept in bounds, other *plants* are preserved free from it, the ground is increased [w],

---

[w] This observation is found in Linn. Flor. Lapp. p. 62, where he says the Dutch sow this grass on their sand banks, that the sand may not overwhelm the neighboring parts. I do not see why this experiment should not be tryed on the barren sands in Norfolk, where i am assured by credible witnesses, that the small cottages are sometimes totally buried under sand during high winds. This grass grows plentifully along the sea shores in England. Vid. Ray, 393. 1.

and the sea repelled by this wonderful difposition of nature.

How folicitous nature is about the prefervation of graffes is abundantly evident from hence, that the more the leaves of the perennial graffes are eaten, the more they creep by the roots, and fend forth off-fets. For the Author of nature intended, that *vegetables* of this kind, which have very flender, and erect leaves, fhould be copious, and very thick-fet, covering the ground like a carpet; and thus afford food fufficient for fo vaft a quantity of grazing animals. But what chiefly increafes our wonder is, that although the graffes are the principal food of fuch animals, yet they are forbid, as it were, to touch the flower, and feed-bearing ftems; that fo the feeds may ripen and be fown.

The *caterpillar* or *grub* of the *moth*, Faun. Suec. 826. called *graefmafken*, although it feeds upon graffes, to the great deftruction of them, in meadows; yet it feems to be formed, in order to keep a due proportion between thefe and other plants; for graffes when left to grow freely, increafe to that degree, that they exclude all other plants; which would confequently be extirpated, unlefs this infect fome-

times

OF NATURE.

times prepared a place for them. Hence always more species of *plants* appear in those places where this caterpillar has laid waste the pastures the preceding year, than at any other time.

§. 10.

Destruction.

Daily experience teaches us, that all *plants* as well as all other living things, must submit to death.

They spring up, they grow, they florish, they ripen their fruit, they wither, and at last, having finished their course, they die, and return to the dust again, from whence they first took their rise. Thus all black mould, which every where covers the earth, for the greatest part is owing to dead *vegetables*. For all roots descend into the sand by their branches, and after a *plant* has lost its stem the root remains; but this too rots at last, and changes into mould. By this means this kind of earth is mixed with sand, by the contrivance of nature, nearly in the same way as dung thrown upon fields is wrought into the earth by the industry of the husbandman. The earth thus prepared offers again to *plants* from its bosom, what it has received from them. For when seeds are committed to the earth, they draw to themselves, accommodate to their nature, and turn into *plants*, the more subtile parts of this mould by the co-operation of the sun, air, clouds, rains, and winds; so that the tallest tree is, properly speaking, nothing but mould wonderfully compounded with air, and water, and modified by a vertue communicated to a small seed by the Creator. From these plants, when they die, just the same kind of mould is formed, as gave birth to them originally; but in such a manner, that it is in greater quantity than before. *Vegetables* therefore increase the black mould, whence fertility remains continually uninterrupted. Whereas the earth could not make good its annual consumption, unlefs it were constantly recruited by new supplies.

The crustaceous *liverworts* are the first foundation of *vegetation*, and therefore are *plants* of the utmost consequence in the œconomy of nature, though so despised by us. When rocks first emerge out of the sea, they are so polished by the force of the waves, that
scarce

scarce any herb can find a fixed habitation upon them; as we may observe every where near the sea. But the very minute crustaceous *liverworts* begin soon to cover these dry rocks, although they have no other nourishment, but that small quantity of mould, and imperceptible particles, which the rain and air bring thither. These *liverworts* dying at last turn into a very fine earth; on this earth the * imbricated *liverworts* find a bed to strike their roots in. These also dye after a time, and turn to mould; and then the various kinds of mosses, e. g. the *hypna*, the *brya*, *politricha* find a proper place, and nourishment. Lastly, these dying in their turn, and rotting afford such a plenty of new formed mould, that herbs and shrubs easily root, and live upon it.

That trees when they are dry or are cut down may not remain useless to the world, and lye, as it were, melancholy spectacles, nature hastens on their destruction in a singular way: first the *liverworts* begin to strike root in them; afterwards the moisture is drawn out of them;

---

* I have used this word because we have no English one of the same meaning unless it be the word *scaly*, that I know of. However imbricated means parts lying over parts like tiles, as in the cup of the *thistle flower*.

whence

whence putrefaction follows. Then the *mush-room* kinds find a fit place for nourishment on them, and corrupt them still more. The *beetle* called the *dermestes*, next makes himself a way between the bark and the wood. The *musk-beetle*, the *copper tale beetle*, and the *caterpillar* or *cossus* 812. bores an infinite number of holes through the trunk. Lastly the *woodpeckers* come, and while they are seeking for insects, wear away the tree, already corrupted; till the whole passes into earth. Such industry does nature use to destroy the trunk of a tree! Nay trees immersed in water would scarcely ever be destroyed, were it not for the worm that eats ships, which performs this work; as the sailor knows by sad experience.

Thistles, as the most usefull of plants, are armed and guarded by nature herself. Suppose there were a heap of clay, on which for many years no plant has sprung up; let the seeds of the *thistle* blow there, and grow, the *thistles* by their leaves attract the moisture out of the air, send it into the clay by means of their roots, will thrive themselves, and afford a shade. Let now other plants come hither, and they will soon cover the ground. St. Bielķe.

All

All succulent plants make ground fine, of a good quality, and in great plenty, as *sedum, crassula, aloe, algæ* [y]. But dry plants make it more barren, as *ling* or *heath, pines, moss*; and therefore nature has placed the succulent plants on rocks, and the dryest hills.

§. 11.

The animal kingdom.

Propagation.

The generation of animals holds the first place among all things, that raise our admiration, when we consider the works of the Creator; and that appointment particularly, by which he has regulated the conception of the *fœtus*, and its exclusion, that it should be adapted to the disposition, and way of living of each animal, is most worthy of our attention.

We find no species of animals exempt from the stings of love, which is put into them to the end, that the Creator's mandate may be executed, *increase and multiply*; and that thus

[y] A kind of *grass wrack*.

the egg, in which is contained the rudiment of the *fœtus* may be fœcundated; for without fœcundation all eggs are unfit to produce an offspring.

Foxes and *wolves*, struck with these stings, every where howl in the woods; crowds of *dogs* follow the female: *bulls* shew a terrible countenanee, and very different from that of *oxen*. *Stags* every year have new horns, which they lose after rutting time. Birds look more beautifull than ordinary, and warble all day long through lasciviousness. Thus *small birds* labour to outsing one another, and *cocks* to outcrow. *Peacocks* spread forth again their gay, and glorious trains. *Fishes* gather together, and exult in the water; and *grashoppers* chirp, and pipe as it were, amongst the herbs. The *ants* gather again into colonies, and repair to their citadels[*]. I pass over many other particulars, which this subject affords, to avoid prolixity.

[*] See this subject treated with great spirit in Thomson's Spring and in Virgil's Georgics.

§, 12.

§. 12.

The fœcundated egg requires a certain, and proportionate degree of heat for the expansion of the *stamina* of the *embryo*. That this may be obtained, nature operates in different manners, and therefore we find in different classes of animals a different way of excluding the *fœtus*.

The females of *quadrupeds* have an *uterus*, contrived for easy gestation, temperate and cherishing warmth, and proper nourishment of the *fœtus*, as most of them live upon the earth, and are there fed.

Birds, in order to get subsistence, and for other reasons, are under a necessity of shifting place: and that not upon their feet but wings. Gestation therefore would be burthensome to them. For this reason they lay eggs, covered with a hard shell. These they sit upon by a natural instinct, and cherish till the young one comes forth.

The *ostrich* and *cassowary* are almost the only birds, that do not observe this law; these commit their eggs to the sand, where the intense heat of the sun excludes the *fœtus*.

Fishes inhabit cold waters, and most of them have cold blood; whence it happens that they have not heat sufficient to produce the *fœtus*. The all-wise Creator therefore has ordained, that most of them should lay their eggs upon the shore; where, by means of the solar rays, the water is warmer, and also fitter for that purpose; because it is there less impregnated with salt, and consequently milder; and also because water-insects abound more there, which afford the young fry a nourishment.

Salmons in the like manner, when they are about to lay their eggs, are led by instinct to go up the stream, where the water is fresh and more pure.

The *butterfly fish* is an exception, for that brings forth its *fœtus* alive.

The *fishes of the ocean*, which cannot reach the shores by reason of the distance, are also exempt from this law. The Author of nature to this kind has given eggs that swim: so that they are hatched amidst the swimming *fucus*, called *sargazo*. Flor. Zeilon. 389.

The *cetaceous* fish have warm blood, and therefore they bring forth their young alive, and suckle them with their teats.

Many

OF NATURE.

Many *amphibious* animals bring forth live *fœtuses*. As the *viper*, and the *toad*, &c. But the species that lay eggs, lay them in places, where the heat of the sun supplies the warmth of the parent.

Thus the rest of the *frog* kind and the *lizard* kind, lay their eggs in warm waters; the common *snake* in dunghills,' and such-like warm places, and give them up to nature, as a provident nurse to take care of them. The *crocodile*, and *sea tortoises* go ashore to lay their eggs under the sand, where the heat of the sun hatches them.

Most of the *insect* kind neither bear young nor hatch eggs; yet their tribes are the most numerous of all living creatures; insomuch that if the bulk of their bodies were proportionate to their quantity, they would scarce leave room for any other kind of animals. Let us see therefore with what wisdom the Creator has managed about the propagation of these minute creatures. The females by natural instinct meet and copulate with the males; and afterwards lay their eggs, but not indiscriminately in every place; for they all know how to choose such places as may supply their offspring in its tender age with nourishment, and

other

other things necessary to satisfy their natural wants; for the mother, soon after she has laid her eggs, dyes, and were she to live she would not have it in her power to take care of her young.

Butterflies, moths, some *beetles, wevils, bugs, cuckow-spit insects, gall-insects, tree bugs,* &c. lay their eggs on the leaves of plants, and every different tribe chooses its own species of plant *. Nay there is scarce any plant, which does not afford nourishment to some insect; and still more, there is scarcely any part of a plant, which is not preferred by some of them. Thus one insect feeds upon the flower; another upon the trunk: another upon the root; and another upon the leaves. But we cannot help wondering particularly, when we see how the leaves of some trees, and plants, after eggs have been let into them, grow into galls; and form dwellings, as it were, for the young ones, where they may conveniently live. Thus when the *gall-insect* called *cynips,* Fn. 947. has fixed her eggs in the leaves of an oak, the wound of the leaf swells, and a knob like an apple arises, which includes and nourishes the embryo.

* Vid. Syst. Nat. Edit. 10. Fauna Suecica; and Hospita Insectorum Flora Amæn. Academ. vol. 3.

OF NATURE. 87

When the *tree-bug*, Faun. Suec. 700. has deposited its eggs in the boughs of the fir tree, excrescences arise shaped like peas. When another species of the *tree-bug*, Fn. 695. has deposited its eggs in the *mouse-ear chickweed* or the *speedwell*, Fl. 12. the leaves contract in a wonderfull manner into the shape of a head. The *water-spider*, Fn. 1150. excludes its eggs either on the extremities of the *juniper*, which from thence forms a lodging, that looks like the *arrow-headed grass*, or on the leaves of the *poplar*, from whence a red globe is produced. The *tree-louse*, Fn. 1355. lays its eggs on the leaves of black poplar, Fl. 821. which from thence turn into a kind of inflated bag, and so in other instances. Nor is it upon plants only that insects live, and lay their eggs. The *knats*, Fn. 1116. commit theirs to stagnating waters. The water insect called *monoculus*, Fn. 1182. often increases so immensely on pools, that the red legions of them have the appearance of blood. Others lay their eggs in other places, e. g. the *beetle* in dunghills. The *dermestes* in skins. The *flesh fly* in putrified flesh. The *cheese-maggot* in the cracks of cheese, from whence the *caterpillars* issuing forth oftentimes consume the whole cheese, and deceive many people, who fancy the worms are produced from the particles of the cheese itself, by a generation called æquivocal, which is extremely absurd. Others exclude their eggs upon certain animals. The *mill-beetle* Fn. 618. lays its eggs between the scales of fishes. The *species of glad-fly* Fn. 1024 on the back of cattle. The *species* 1025 on the back of the rhen deer. The *species* 1026 in the noses of sheep. The *species* 1028 lodges during the winter in the intestinal tube, or the throat of horses, nor can it be driven out till the summer comes on. Nay *insects* themselves are often surrounded with the eggs of other insects, insomuch that there is scarcely an animal to be found, which does not feed its proper insect, not to say any more of all the other places where they deposit their eggs. Almost all the eggs of *insects*, when laid, are ordained to undergo, by a wonderfull law of nature, various metamorphoses, e. g. the egg of the *butterfly* being laid in the cabbage first of all becomes a *caterpillar*, that feeds upon the plant, crawls, and has sixteen feet. This afterwards changes into a *nymph* that has no feet, is smooth, and eats nothing; and lastly this bursts into a *butterfly*, that flies, has variety. of colours, is rough,

and

OF NATURE.

and lives upon honey. What can be more worthy of admiration, than that one, and the same animal should appear on the stage of life under so many characters, as if it were three distinct animals [a].

The laws of generation of *worms* are still very obscure, as we find they are sometimes produced by eggs, sometimes by offsets, just in the same manner as happens to trees. It has been observed with the greatest admiration, that the *polypus* or *hydra* S. N. 221. lets down shoots and live branches, by which it is multiplied. Nay more, if it be cut into many parts, each segment, put into the water, grows into a perfect animal; so that the parts which were torn off are restored from one scrap.

§. 13.

The multiplication of animals is not tyed down to the same rules in all; for some have a remarkable power of propagating, others are

[a] Linnæus Amæn. academ. vol. 2. in a treatise on the wonders relating to insects, says, "as surprising as these transformations may seem, yet much the same happens when a chicken is hatched; the only difference is, that the chicken breaks all three coats at once, the butterfly one after another."

confined within narrower limits in this respect. Yet in general, we find, that nature observes this order, that the least animals, and those which are usefull, and serve for nourishment to the greatest number of other animals, are endued with a greater power of propagating than others [b].

Mites, and many other insects will multiply to a thousand within the compass of a very few days. While the *elephant* scarcely produces one young in two years.

The *hawk* kind generally lay not above two eggs, at most four, while the *poultry* kind rise to 50.

The *diver* or *loon*, which is eaten by few animals, lays also two eggs, but the *duck* kind, the *moor game*, *partridges*, &c. and *small birds* lay a very large number.

If you suppose two *pigeons* to hatch nine times a year, they may produce in four years 14672 young [c]. They are endued with this

[b] Herodotus speaking of the flying serpents in Arabia makes the same reflection, and attributes this course of nature to the divine providence. Thal.

[c] I have given this passage as it stands in the original. The numbers ought to have been 14760, or the expression should have been altered; for he includes the first pair.

He supposes it generally known that pigeons hatch but two eggs at a time, and that they pair.

remark-

OF NATURE. 91

remarkable fertility, that they may serve for food, not only to man, but to hawks and other birds of prey*. Nature has made harmless and esculent animals fruitfull. Plin. Nature has forbid the *bird* kind to fall short of the number of eggs allotted to each species, and therefore if the eggs which they intend to sit upon, be taken away a certain number of times, they presently lay others in their room, as may be seen in the swallow, duck, and small birds.

§. 14.
Preservation.

Preservation follows generation; this appears chiefly in the tender age, while the young are unable to provide for their own support. For then the parents, though otherwise ever so fierce in their disposition, are affected with a wonderfull tenderness or sense of love towards their progeny, and spare no pains to provide for, guard, and preserve them, and that not by an imaginary law, but one given by the Lord of nature himself.

Quadrupeds give suck to their tender young, and support them by a liquor, perfectly easy of digestion, till their stomachs are able to digest,

* Vid. Muschenbr. Orat. de Sap. Divin.

and

and their teeth are fit to chew more solid food. Nay their love toward them is so great, that they endeavour to repell with the utmost force every thing, which threatens danger, or destruction to them. The *ewe* which brings forth two *lambs* at a time, will not admit one to her teats, unless the other be present, and suck also; left one should famish, while the other grows fat.

Birds build their nests in the most artificial manner, and line them as soft as possible, for fear the eggs should get any damage. Nor do they build promiscuously in any place; but there only, where they may quietly lye concealed, and be safe from the attacks of their enemies.

The *hanging bird*, Act. Bonon. vol. 2. makes its nest of the fibres of withered plants, and the down of the poplar seeds, and fixes it upon the bough of some tree hanging over the water, that it may be out of reach.

The *diver*, Fn. 123. places its swimming nest upon the water itself amongst the rushes. I designedly pass over many other instances of the like kind.

Again birds sit on their eggs with so much patience, that many of them choose to perish
with

OF NATURE.

with hunger, rather than expose the eggs to danger by going to seek for food.

The male *rooks* and *crows* at the time of incubation bring food to the females.

Pigeons, small birds, and other *birds,* which pair, sit by turns; but where polygamy prevails, the males scarcely take any care of the young.

Most of the *duck* kind pluck off their feathers in great quantity, and cover their eggs with them, lest they should be damaged by the cold, when they quit their nests for the sake of food; and when the young are hatched, who knows not how solicitous they are in providing for them, till they are able to fly and shift for themselves?

Young *pigeons* would not be able to make use of hard seeds for nourishment, unless the parents were to prepare them in their crops, and thence feed them.

The *eagle owl* makes its nest on the highest precipices of mountains, and in the warmest spot, facing the sun; that the dead bodies brought there may by the heat melt into a soft pulp, and become fit nourishment for the young.

The

The *cuckow* lays its eggs in the nest of other small birds, generally the *wagtail*, [d] or [e] *hedge-sparrow*, and leaves the incubation, and preservation of the young to them. But that these young, when grown up, degenerate into hawks, and become so ungratefull, that they destroy their nurses, is a mere vulgar error, for it is contrary to their nature to eat flesh.

Amphibious animals, *fishes* and *insects*, which cannot come under the care of their parents, yet owe this to them, that they are put in places, where they easily find nourishment, as we have observed.

[d] This custom of the cuckow is so extraordinary, and out of the common course of nature, that it would not be credible, were it not for the testimony of the most knowing and curious natural historians, such as Ray, Willughby, Gesner, Aldrovandus, Aristotle, &c.

Much has been said by the writers on birds about the fate of the young birds, in whose nest the cuckow is hatched, but as i find nothing but mere conjecture, it would not be worth while transcribing.

[e] Hedge-sparrow. Linnæus seems to have taken the white-throat for the hedge-sparrow.

OF NATURE.

§. 15.

As soon as animals come to maturity, and want no longer the care of their parents, they attend with the utmost labour, and industry, according to the law and œconomy appointed for every species, to the preservation of their lives. But that so great a number of them, which occur every where, may be supported, and a certain and fixed order may be kept up amongst them, behold the wonderful disposition of the Creator, in assigning to each species certain kinds of food, and in putting limits to their appetites. So that some live on particular species of plants, which particular regions, and soils only produce. Some on particular animalcula, others on carcases, and some even on mud and dung. For this reason Providence has ordained, that some should swim in certain regions of the watery element, others should fly; some should inhabit the torrid, the frigid, or the temperate zones, and others should frequent desarts, mountains, woods, pools or meadows, according as the food proper to their nature is found in sufficient quantity. By this means there is no terrestrial
tract,

tract, no sea, no river, no countrey, but what contains, and nourishes various kinds of animals. Hence also an animal of one kind cannot rob those of another kind of its aliment; which, if it happened, would endanger their lives or health; and thus the world at all times affords nourishment to so many, and so large inhabitants, at the same time that nothing which it produces, is useless or superfluous.

I think it will not be amiss to produce some instances, by which it will appear, how providentially the Creator has furnished every animal with such cloathing, as is proper for the countrey where they live, and also how excellently the structure of their bodies is adapted to their particular way of life; so that they seem to be destined solely to the places, where they are found.

Monkies, elephants, and *rhinoceroses* feed upon vegetables, that grow in hot countries, and therefore therein they have their allotted places. When the sun darts forth its most fervid rays, these animals are of such a nature, and disposition, that it does them no manner of hurt; nay with the rest of the inhabitants of those parts they go naked, whereas were

OF NATURE.

they covered with hairy skins, they must perish with heat.

On the contrary the place of the *rhen deer* is fixed in the coldest part of Lapland, because their chief food is the *liverwort*, Fl. 980. which grows no where so abundantly as there; and where, as the cold is most intense, the *rhen deer* are cloathed, like the other northern animals, with skins filled with the densest hair; by the help of which they easily defy the keenness of the winter. In like manner the *rough-legged partridge* passes its life in the very Lapland alps, feeding upon the seeds of the *dwarf birch*, and that they may run up and down safely amidst the snow, their feet are feathered.

The *camel* frequents the sandy, and burning desarts, in order to get the barren *camel's hay*. Mat. Med. 31. How wisely has the Creator contrived for him! he is obliged to go thro' the desarts, where oftentimes no water is found for many miles about. All other animals would perish with thirst in such a journey; but the camel can undergo it without suffering; for his belly is full of cells, where he reserves water for many days. It is reported by travellers, that the Arabians, when in travelling they want water, are forced to kill their camels, and take water out of their bellies, that is perfectly good to drink, and not at all corrupted.

The *pelican* likewise lives in defart, and dry places; and is obliged to build her nest far from the sea, in order to procure a greater share of heat to her eggs. She is therefore forced to bring water from afar for herself and her young; for which reason Providence has furnished her with an instrument most adapted to this purpose; v. g. she has a very large bag under her throat, which she fills with a quantity of water sufficient for many days; and this she pours into the nest to refresh her young, and teach them to swim. The wild beasts, lions, and tigers, come to this nest to quench their thirst, but do no hurt to the young.

Oxen delight in low grounds, because there the food most palatable to them grows.

Sheep prefer naked hills, where they find a particular kind of grafs called the *festuca*, Fl. 95. which they love above all things.

Goats climb up the precipices of mountains, that they may browse on the tender shrubs, and in order to fit them for it, they have feet made for jumping*.

* Vid. Derham's Physico-Theol. p. 319. not. 7.

Horses

OF NATURE.

Horses chiefly resort to woods, and feed upon leafy plants.

Nay, so various is the appetite of animals, that there is scarcely any plant, which is not chosen by some, and left untouched by others. The *horse* gives up the *water hemlock* to the goat. The *cow* gives up the *long-leaved water hemlock* to the sheep. The *goat* gives up the *monks-hood* to the horse, &c. for that which certain animals grow fat upon, others abhor as poison. Hence no plant is absolutely poisonous, but only respectively. Thus the *spurge*, that is noxious to man, is a most wholesome nourishment to the *caterpillar*, Fn. 825. That animals may not destroy themselves for want of knowing this law, each of them is guarded by such a delicacy of taste and smell, that they can easily distinguish what is pernicious from what is wholesome; and when it happens that different animals live upon the same plants, still one kind always leaves something for the other, as the mouths of all are not equally adapted to lay hold of the grass; by which means there is sufficient food for all. To this may be referred an œconomical experiment well known to the Dutch, that when eight cows have been in a pasture, and can no longer

longer get nourishment, two horses will do very well there for some days, and when nothing is left for the horses, four sheep will live upon it.

Swine get provision by turning up the earth; for there they find the succulent roots, which to them are very delicious.

The leaves and fruits of trees are intended as food for some animals, as the sloth [f], the

---

[f] There is so curious an account of this animal in Kircher's Musurgia, that I think the reader will excuse my transcribing it. That author says thus: ' The description
' of this animal i had from father Torus, provincial of the
' Jesuites in America, who had animals of this kind in
' his possession, and made many experiments in relation to
' their nature and qualities. Its figure is extraordinary;
' it is about the bigness of a cat, of very ugly countenance,
' and has claws extended like fingers. The hinder part
' of the head and neck are covered with hair. It sweeps
' the ground with its fat belly, never rises upon its feet,
' and moves so slowly, that it would scarce go the length
' of a bow-shot in 15 days, tho' constantly moving, and it
' is therefore called the Sloth. It is not known what it
' feeds upon, not being ever observed to take any food. It
' lives generally upon tops of trees, and employs two days
' to crawl up and as many to get down again. Nature has
' doubly guarded this animal against its enemies. First by
' giving it such strength in its feet that whatever it seizes,
' it holds so fast, that it can never be freed from its claws,
' but must there die of hunger. Secondly in giving it such
' a moving

OF NATURE.

the squirrel, and these last have feet given them fit for climbing.

Besides myriads of fishes, the *castor*, the *sea calf*, and others inhabit the water, that they may there be fed, and their hinder feet are fit for

'a moving aspect, when it looks at any man who should
'be tempted to hurt it, that it is impossible not to be
'touched with compassion; besides that at the same time
'it sheds tears, and upon the whole persuades one that a
'creature so defenceless, and of so unhappy a body ought
'not to be tormented. To make an experiment of this,
'the abovementioned father procured one of these animals
'to be brought to our college at Carthagena. He put a
'long pole under his feet, which it seized upon very firmly
'and would not let it go again. The animal therefore
'thus voluntarily suspended was placed between two beams
'along with the pole, and there it remained without meat,
'drink, or sleep, forty days; its eyes being always fixed
'on people that looked at it, who were so touched, that
'they could not forbear pitying it. At last being taken
'down they let loose a dog on it, which after a little while
'the Sloth seized with his feet, and held him four
'days, till he died of hunger. This was taken from the
'mouth of the father. They add, continues Kircher,
'that this creature makes no noise but at night, but that
'very extraordinary. For by interruptions, that last a-
'bout the length of a sigh or semipause, it goes through
'the six vulgar intervals of music, ut, re, mi, fa, sol, la,
'La, sol, fa, mi, re, ut, ascending and descending, and
'these perfectly in tune. So that the Spaniards, when
'they first got possession of this coast, and heard these

'notes

for swimming, and perfectly adapted to their manner of life.

The whole order of the *goose* kind, as ducks, merganser, &c. pass their lives in water, as feeding upon water-insects, fishes, and their eggs ᶠ. Who does not see, that attends ever so little, how exactly the wonderfull for-

'notes, imagined that some people brought up to our
'music, were singing. This animal is called by the
'natives, Haut, certainly because going thro' these mu-
'sical intervals, it repeats, Ha, ha, ha, ha, ha, &c.'

This account seems very wonderfull, and i leave it as it stands without entering into any discussion about its credibility. I will only add, that Linnæus seems in the new edition of the Syst. Nat. to give credit to it. For he says in his short way of description among other things, 'It utters an ascending hexacord. Its noise is horrible, 'its tears piteous.' He quotes Mangrave, Clusius, Gesner, &c. But not having an opportunity of consulting these books, i cannot tell how far these authors confirm the foregoing account; if it be true, it would furnish some observations, but this would not be a place for them.

ᶠ Many opinions, says the author in the note, have been started in order to account how it happens that fishes are found in pools, and ditches, on high mountains and elsewhere. But Gmelin observes that the *duck* kind swallow the eggs of fishes, that some of these eggs go down, and come out of their bodies unhurt, and so are propagated just in the same manner, as has been observed of plants. Biberg.

Gmelin adds, that the Sibirians themselves account for this phænomenon in the manner above mentioned.

mation

OF NATURE. 103

mation of their beaks, their necks, their feet, and their feathers suit their kind of life, which observation ought to be extended to all other birds.

The way of living of the *sea-swallow* Fn. 129. deserves to be particularly taken notice of; for as he cannot so commodiously plunge into the water and catch fish as other aquatic birds, the Creator has appointed the *sea-gull* to be his caterer in the following manner. When this last is pursued by the former, he is forced to throw up part of his prey, which the other catches; but in the autumn, when the fishes hide themselves in deep places, the merganser, Fn. 113. supplies the gull with food, as being able to plunge deeper into the sea. Act. Stock.

The chief granary of *small birds* is the *knot-grass*, Fol. Suec. 322. that bears heavy seeds, like those of the *black bindweed*. It is a very common plant, not easily destroyed, either by the road side by trampling upon it, or any where else, and is extremely plentifull after harvest in fields, to which it gives a reddish hue, by its numerous seeds. These fall upon the ground, and are gathered all the year round by the small birds.

ʰ Thus bountifull nature feeds the fowls of the air.

The Creator has taken no lefs care of some *amphibious animals*, as the snake and frog kind, which, as they have neither wings to fly, nor feet to run swiftly, and commodiously, would scarcely have any means of taking their prey, were it not that some animals run, as it were of their own accord, into their mouths. When the *rattle-snake*, a native of America, with open jaws fixes his eyes on a bird, fly, or squirrel, sitting on a tree, they fly down his throat, being rendered stupid, and giving themselves up, as destitute of all refuge. On the other hand we cannot but adore the Creator's great goodness towards man, when we

ʰ To which we may add, that many small birds feed upon the seeds of *plantain*, particularly linnets. It is generally known, that the goldfinch lives upon the seed of *thistles*, from which he has its name in Greek, Latin, and French.

ⁱ How dreadful this serpent is to other animals will appear by an account we have in a treatife intitled. Radix Senega. Where the author Amœn. academ. vol 2, says, one of their terrible serpents got clandestinely into the house of governor Blake at Cacolina; where it would have long laid concealed, had it not been that all the domestic animals, as dogs, hogs, turkies and fowls admonished the family by their unusual cries, equally shewing their horror and consternation, their hair, bristles, and crests standing up an end.

consider

consider the rattle which terminates this serpent's tail. For by the means of that we have an opportunity of guarding againſt this dreadfull enemy; the ſound warning us to fly, which if we were not to do, and we ſhould be wounded by him, the whole body would be turned into a putrid corruption in ſix hours, nay ſometimes in half an hour.

The limits of this diſſertation will not permit me to produce more examples of this kind. But whoever will be at the pains to take ever ſo ſlight a view of the wonderfull works of the Creator, will readily ſee how wiſely the plan, order and fitneſs of things to divine ends are diſpoſed.

§. 16.

We cannot without the utmoſt admiration behold how providently the Creator has acted as to the preſervation of thoſe animals, which at a certain time of the year, are by the rigor of the ſeaſon excluded from the neceſſaries of life. Thus the *bear* in the autumn creeps into the *moſs*, which he has gathered, and there lies all winter; ſubſiſting upon no other nouriſhment but his fat, collected during the ſummer in the cellulous membrane, and which without

without doubt, during his fast, circulates thro' his vessels, and supplies the place of food; to which perhaps is added that fat juice which he sucks out of the bottom of his feet.

The *hedge-hog*, *badger* and *mole* in the same manner fill their winter quarters with vegetables, and sleep during the frosts.

The *bat* seems cold, and quite dead all the winter. Most of the *amphibious animals* get into dens, or to the bottom of lakes and pools.

In the autumn, as the cold approaches, and insects disappear, *swallows*[k] seek for an asylum against the violence of the cold in the bottom
of

---

[k] I never had but one credible testimony that swallows pass the winter at the bottom of lakes or ponds; and this from a gentleman of character, who saw a swallow so found brought to life by warmth. On the other hand, i know of no author but Herodotus who mentions their being seen in any countrey during the winter. He, lib. 2, p. 109. edit. Steph. says, that swallows and kites continue all the year about the springs of the Nile. What he mentions concerning kites deserves some notice, viz. that they lye concealed in holes a few days. Pliny says a few months. Gesner repeats the same, adding that they have been found in hollow trees somewhere in Upper Germany, but he seems to relate this upon hearsay only. Aldrovandus gives the same account as Gesner, and adds that they winter in Egypt, but whether upon the authority of Bellonius or any other credible writer, does not appear.

OF NATURE.

of lakes amongst the reeds and rushes; from whence, by the wonderfull appointment of nature they come forth again. The periſtaltic motion of the bowels ceaſes in all theſe animals, while they are obliged to faſt, whence the appetite is diminiſhed, and ſo they ſuffer leſs from hunger. To this head may be referred the obſervation of the celebrated Liſter concerning thoſe animals; that their blood, when let into a baſon, does not coagulate, as that of all other animals, and ſo is no leſs fit for circulation than before.

The *moor-fowls* work themſelves out walks under the very ſnow. They moult in the ſummer, ſo that about the month of Auguſt they

appear. He quotes a paſſage from that author concerning the appearance of a vaſt number of kites at the mouth of the Boſphorus, but this happened at the latter end of May, and ſeems to prove nothing; for the time marked for their appearance by Calippus, who obſerved near the Helleſpont, is the month of March. Willughby ſays that kites are ſuppoſed to be birds of paſſage, and then quotes from Bellonius the place abovementioned.

From what has been ſaid it appears evident, that nothing certain is known by the moderns about the diſappearance of theſe remarkable birds, yet their coming was regularly noted by the antient writers, and coincided with that of ſwallows, as appears by the old calendars of Geminus and Ptolemy from the obſervations of Eudoxus, Euctemous, Calippus, and Doſitheüs.

cannot fly, and are therefore obliged to run into the woods; but then the moor-berries, and bilberries are ripe, from whence they are abundantly supplied with food. Whereas the young do not moult the first summer, and therefore tho' they cannot run so well, are able to escape danger by flight.

The *rest of the birds* who feed upon insects migrate every year to forreign regions, in order to seek for food in a milder climate; while all the northern parts, where they live well in the summer, are covered with snow.

Insects in the winter generally lye hid within their cases, and are nourished by the surrounding liquor, like the fœtus of other animals, from whence at the approach of spring they awake, and fly forth to the astonishment of every one.

However all animals which lye hid in winter, do not observe these laws of fasting. Some provide store-houses in summer, and autumn, from which they take what is necessary, as *mice, jays, squirrels, bees*.

§. 17.

What i have observed in a few words concerning the migration of birds into forreign coun-

OF NATURE. 109

countries, gives me an opportunity of illuftrating this fubject farther by inftances.

The *ftarling*, Fn. 183. finding with us after the middle of fummer worms in lefs plenty, yearly goes into Schonen, Germany and Denmark.

The *female chaffinches* every winter, about Michaelmas, go in flocks to Holland; but as the males ftay with us, they come back the next fpring, unlefs fuch as choofe to breed no more.

In the fame manner the female *Caroline yellow-hammer* in the month of September, while the rice, on which fhe feeds, is laid up in granaries, goes towards the fouth, and returns in the fpring to feek her mate.

Our *aquatic birds* are forced by neceffity to fly towards the fouth every autumn before the water is frozen. Thus we know that the lakes of Poland and Lithuania are filled with *fwans* and *geefe* every autumn, at which time they go in great flocks along many rivers as far as the Euxine. But in the beginning of fpring, as foon as the heat of the fun molefts them, they turn back, and go again to the northern pools, and lakes, in order to lay their eggs. For there, and efpecially in Lapland, there is a vaft abundance of knats Fn. 1116. which afford them

excellent

excellent nourishment, as all of this kind live in the water, before they get their wings.

The *woodcock* Fn. 141. lives in England in winter, and departs from thence at the coming on of spring after they have paired.

The *swallow-tail'd sheldrake* Fn. 96. crosses Sweden in April, and does not stop till she has reached the White sea.

The *coblers awl* Fn. 137. goes every autumn into Italy.

The *arctic driver* Fn. 121. goes into Germany every spring and autumn.

The *missel thrush* Fn. 189. fills our woods in the spring, but leaves us in the winter.

The *pied chaffinch* Syst. Nat. 10. 97. 1. during the winter, being obliged to leave the alps*, hastens into Sweden, and often into Germany.

The *gulls* visit Spain and Italy.

The *raven* [1] goes into Schonen.

By these migrations birds also become useful to many different countries, and are distributed over almost all the globe. I cannot forbear expressing my admiration here, that all

* The Author means the Northern alps.

[1] I have translated the word corvus by raven, because Linnæus does not mention the carrion crow at all, either in the Faun. Suec. nor in the Syst. Nat. before the late edition.

of

OF NATURE. 111

of them exactly observe the times of coming and going, and that they do not mistake their way.

There is a very large shell-fish in the Mediterranean called the *pinna*, blind as all of that genus, but furnished with very strong calcareous valves. (Bell. aquat. 401. t. 401. Jonst. exsang. t. 16. f. 5, 6. Gualt. ind. t. 79, 79.) The *scuttle fish* (Bell. aquat. 330. t. 331. Jonst. exsang. t. 1. f. 1.) is an inhabitant of the same sea, and a deadly enemy to the former; as soon as the *scuttle-fish* sees the *pinna* open its shell, he rushes upon her like a lion, and devours her. The *pinneteres* or *pinnophylax* (Jonst. exsang. t. 20. f. 3.) is of the crab kind naked, like the hermit, and very quick-sighted. This *cancer* or *crab* the *pinna* receives into her covering, and when she opens her valves in quest of food, lets him out to look for prey. During this the *scuttle fish* approaches; the *crab* returns with the utmost speed and anxiety to his hostess, who being thus warned of the danger shuts her doors, and keeps out the enemy. That very sagacious observer D. D. Hasselquist in his voyage towards Palestine beheld this curious phænomenon, which tho' well known to the antients had escaped the moderns. Arist. hist. lib. 5. c. 15. relates,

that

that the *pinna* kept a guard to watch for her. That there grew to the mouth of the *pinna* a small animal, having claws, and serving as a caterer, which was like a *crab*, and was called the *pinnophylax*. Plin. lib. 9. 51. says, the smallest of all the kinds is called the *pinnoteres*, and therefore liable to injury; this has the prudence to hide itself in the shells of *oysters*. Again lib. 9. 66. he says ᵐ the *pinna* is of the genus of shell-fish; it is produced in muddy waters, always erect, nor ever without a companion, which some call the *pinnoteres*, others the *pinnophylax*. This sometimes is a small *squill*, sometimes a *crab*, that follows the *pinna* for the sake of food. The *pinna* is blind, and

---

ᵐ This is taken out of Aristotle, who seems to have thought, that the pinna grew from that which really is its beard, and which it throws out upon the adjoyning bodies in order to fix itself. For he says the pinna is produced from the byssus, which is generally supposed to mean the beard of this shell-fish, and to have been used for making the finest of stuffs, frequently mentioned by antient writers under the name of Byssine garments, and of which they now in some countries make stockings as i am informed. This notion of the pinna growing from the byssus or beard is of the same kind with that which prevailed formerly in relation to the goose tree, mentioned by many writers, of whom a long list may be seen in the tenth edition of the Syst. Nat.

when

when upon opening its shell it exposes itself as a prey to the smallest kind of fishes, these immediately assault her, and growing bolder upon finding no resistance venture in. The guard watching its time gives notice by a bite; upon which the *pinna* closing its shell, shuts in, kills, and gives part of whatever happens to be there to its companion.

The *pinna*, and the *crab* together dwell,
For mutual succour in one common shell.
They both to gain a livelihood combine;
That takes the prey, when this has given the sign.
From hence this *crab* above his fellows famed,
By antient Greeks was *pinnoteres* named.

OPPIAN.

§. 18.

Destruction.

We have observed above that all animals do not live upon vegetables, but that there are some which feed upon certain animalcula. Nay there are some which subsist only by rapine, and daily destroy numbers of the peaceable kind.

These animals are destroyed, but in such a manner

manner that the weaker generally are infested by the stronger in a continued series. Thus the *tree-louse* lives upon plants. The fly called *musca aphidivora* lives upon the *tree-louse*. The *hornet* and *wasp fly* upon the *musca aphidivora*. The *dragon fly* upon the *hornet* and *wasp fly*. The *spider* on the *dragon fly*. The *small birds* on the *spider*. And lastly, the *hawk* kind on the *small birds*.

In like manner the [n] *monoculus* delights in putrid waters, the *knat* eats the *monoculus*, the *frog* eats the *knat*, the *pike* eats the *frog*, the *sea calf* eats the *pike*.

The *bat* and *goat-sucker* make their excursions only at night, that they may catch the *moths*, which at that time fly about in vast quantities.

The *wood-pecker* pulls out the *insects* which lie hid in the trunks of trees.

The *swallow* pursues those which fly about in the open air.

The *mole* pursues the *worms*. The large fishes devour the small. Nay, we scarcely know an animal, which has not some enemy to contend with.

[n] An insect that has no name in English, as far as i can find.

OF NATURE.

Amongst quadrupeds *wild beasts* are most remarkably pernicious, and dangerous to others, as the *hawk* kind among birds. But that they may not, by too atrocious a butchery, destroy whole species; even these are circumscribed within certain bounds. First, as to the most fierce of all, it deserves to be noted how few they are in proportion to other animals. Secondly, the number of them is not equal in all countries. Thus France and England breed no *wolves*, and the northern countries no *tigers* or *lions*. Thirdly, these fierce animals sometimes fall upon, and destroy one another. Thus the *wolf* devours the *fox*. The *dog* infests both the *wolf* and *fox*; nay *wolves* in a body will sometimes venture to surround a *bear*. The *tiger* often kills its own male whelps. *Dogs* are sometimes seized with madness and destroy their fellows, or with the mange destroy themselves.

Lastly, wild beasts seldom arrive at so great an age as animals, which live on vegetables. For they are subject from their alcaline diet to various diseases, which bring them sooner to an end.

But although all animals are infested by their peculiar enemies, yet they are often able to elude their violence by ftratagems and force. Thus the *hare* often confounds the dog by her windings.

When the *bear* attacks *sheep* and cattle, they draw up together for mutual defence. *Horses* joyn heads together, and fight with their heels. *Oxen* joyn tails, and fight with their horns.

Swine get together in herds, and boldly oppose themselves to any attack, fo that they are not eafily overcome; and it is worth while to obferve, that all of them place their young, as lefs able to defend themfelves, in the middle, that they may remain fafe during the battle.

Birds by their different ways of flying oftentimes efcape the *hawk*. If the *pigeon* had the fame way of flying as the *hawk*, fhe would hardly ever efcape his claws [o].

It deferves alfo to be remarked, how much fome animals confult their fafety by night. When *horfes* fleep in woods, one by turns remains awake, and, as it were, keeps watch. When *monkies*, S. N. 2. 10. in Brafil fleep

upon

---

[o] As I have, when opportunities offered, meafured and weighed feveral kinds of birds, i fhall here fubjoyn a table of fome of them with the proportions of the weight to the fail. N. B. By fail i mean the extent of the wings and tail.

I do

OF NATURE.

upon trees, one of them keeps awake, in order to give the sign, when the *tiger* creeps towards them,

I do not pretend to accuracy, and i imagine it will not be expected on a subject of this nature.

	Weight Avoirdupois.		Proportion of square inches to the ounce.
	l.	oz.	
Turkey	8	8	$2\frac{1}{4}$
Pheasant	2	8	$2\frac{1}{4}$
Coot	2	8	$2\frac{1}{2}$
Black cock	2	6	$3\frac{1}{2}$
Puttock	1	14	18
Rook	1	3	$10\frac{1}{4}$
Partridge	1	1	3
Ivy owl	0	15	9
Ring-dove	0	14	10
Woodcock	0	10	6
Small hawk	0	$6\frac{3}{4}$	26
Wood-pecker	0	4	9
Cuckow	0	4	18
Missel bird	0	4	14
Snipe	0	4	$9\frac{1}{4}$
Redshank	0	4	9
Cross bill	0	$1\frac{1}{2}$	$11\frac{3}{4}$
House swallow	0	1	18
House sparrow	0	1	12
Wheat-ear	0	1	14
Linnet	0	$0\frac{1}{2}$	$20\frac{3}{4}$
Black cap	0	$0\frac{1}{2}$	18
Stone smich	0	$0\frac{1}{2}$	25
Beccafigo	0	$0\frac{1}{2}$	24
White throat	0	$0\frac{1}{2}$	17
		grains.	
Long tailed titmouse	0	95	25
Regulus cristatus	0	76	23

them, and in case the guard should be caught asleep, the rest tear him to pieces *. Hence the hunting of rapacious animals is not always succesfull, and they are often obliged to labor a whole day to no purpose. For this reason the Creator has given them such a nature, that they can bear fasting a long time. Thus the *lion* lurks in his den many days without famishing, and the *wolf*, when he has once well satisfied his hunger, can fast many weeks without any difficulty.

It appears by this table that the smaller birds in general have more sail in proportion than the larger of the esculent kind, such as the pheasant, partridge, woodcock, ring-dove, &c. and that it should be so contrived appears reasonable on more accounts than one. First, because small birds living, many of them, amongst shrubs and bushes, are obliged to make short and quick motions in hopping from bough to bough, at which time they always make use of their wings; some of them live chiefly on worms and flies, which are not to be caught without great nimbleness, and frequent gardens and houses, and are more liable to the attacks of cats and other animals. And those which live in open fields are exposed to the hawk, and were they not quick at turning they would scarcely ever escape.

Again the different proportions of the bulk to the surface in large and small birds is to the disadvantage of the latter, on account of the greater proportional resistance of the air, and this wanted some compensation.

More might be added on this subject, but i am afraid most readers will think what i have already said is more than enough.

* Maregraf. Braf. 227. Biberg.

OF NATURE.

If we consider the end for which it pleased the Supreme Being to constitute such an order of nature, that some animals should be, as it were, created only to be miserably butchered by others, it seems that his Providence not only aimed at sustaining, but also keeping a just proportion amongst all the species; and so prevent any one of them increasing too much, to the detriment of men, and other animals. For if it be true, as it is most assuredly, that the surface of the earth can support only a certain number of inhabitants, they must all perish, if the same number were doubled, or tripled. Derh. Phys. Theol. p. 237.

There are some viviparous *flies*, which bring forth 2000 young. These in a little time would fill the air, and like clouds intercept the rays of the sun, unless they were devoured by birds, spiders, and many other animals.

Storks, and *falcons* free Ægypt from *frogs*, which after the inundation of the Nile, cover all the countrey. The same birds also clear Palestine of *mice*. Bellonius on this subject says as follows. "The *storks* come to Ægypt "in such abundance, that the fields and mea-"dows are white with them. Yet the Ægyp-"tians are not displeased with this sight; as

|| Muschenbr.

"*frogs*

"*frogs* are generated in such numbers there,
"that did not the *storks* devour them, they
"would over-run every thing. Besides they
"also catch, and eat *serpents*. Between Belba
"and Gaza the fields of Palestine are often
"desert on account of the abundance of *mice*,
"and *rats*; and were they not destroyed by
"the *falcons*, that come here by instinct, the
"inhabitants could have no harvest."

The *white fox* S. N. 8. 7. is of equal advantage in the Lapland alps; as he destroys the Norway *rats*, Fn. 26. which are generated there in great abundance; and thus hinder them from increasing too much in proportion, which would be the destruction of vegetables.

It is sufficient for us, that nothing is made by Providence in vain, and that whatever is made, is made with supreme wisdom. For it does not beome us to pry too boldly into all the designs of God. Let us not imagine, when these rapacious animals sometimes do us mischief, that the Creator planned the order of nature according to our private principles of œconomy; for the Laplanders have one way of living; the European husbandman another; the Hottentots and savages a third, whereas the stupendous œconomy of the Deity is one
through-

OF NATURE.

throughout the globe, and if Providence does not always calculate exactly according to our way of reckoning, we ought to consider this affair in the same light, as when different seamen wait for a fair wind, every one, with respect to the part he is bound to, who we plainly see cannot all be satisfied.

§. 19.

The whole earth would be overwhelmed with carcases, and stinking bodies, if some animals did not delight to feed upon them. Therefore when an animal dyes, *bears, wolves, foxes, ravens,* &c. do not lose a moment till they have taken all away. But if a *horse*, e. g. dyes near the public road, you will find him, after a few days, swoln, burst, and at last filled with innumerable *grubs* of carnivorous *flies*, by which he is entirely consumed, and removed out of the way, that he may not become a nusance to passengers by his poisonous stench.

When the carcases of fishes are driven upon the shore, the voracious kinds, such as the *thornback*, the *hound fish*, the *conger eel*, &c. gather

about

about and eat them. But becaufe the flux, and reflux foon change the ftate of the fea, they themfelves are often detained in pits, and become a prey to the wild beafts, that frequent the fhores. Thus the earth is not only kept clean from the putrefaction of carcafes, but at the fame time by the œconomy of nature the neceffaries of life are provided for many animals. In the like manner many *infects* at once promote their own good, and that of other animals. Thus *knats* lay their eggs in ftagnant, putrid and ftinking waters, and the *grubs* that arife from thefe eggs clear away all the putrefaction; and this will eafily appear, if any one will make the experiment by filling two veffels with putrid water, leaving the *grubs* in one, and taking them all out of the other. For then he will foon find the water, that is full of *grubs*, pure and without any ftench, while the water that has no *grubs* will continue ftinking.

Lice increafe in a wonderfull manner in the heads of children, that are fcabby, nor are they without their ufe, for they confume the redundant humours.

The *beetle* kind in fummer extract all moift and glutinous matter out of the dung of cattle,

OF NATURE.

so that it becomes like dust, and is spread by the wind over the ground. Were it not for this, the vegetables that lye under the dung, would be so far from thriving, that all that spot would be rendered barren.

As the excrements of *dogs* is of so filthy and septic a nature, that no *insect* will touch them, and therefore they cannot be dispersed by that means, care is taken that these animals should exonerate upon stones, trunks of trees, or some high place, that vegetables may not be hurt by them.

Cats bury their dung. Nothing is so mean nothing so little, in which the wonderful order, and wise disposition of nature does not shine forth.

§. 20.

Lastly, all these treasures of nature so artfully contrived, so wonderfully propagated, so providentially supported throughout her three kingdoms, seem intended by the Creator for the sake of man. Every thing may be made subservient to his use, if not immediately, yet mediately, not so to that of other animals. By the help of reason man tames the fiercest animals,

mals, purfues and catches the fwifteft, nay he is able to reach even thofe which lye hid in the bottom of the fea.

By the help of reafon he increafes the number of vegetables immenfely, and does that by art, which nature, left to herfelf, could fcarcely effect. By ingenuity, he obtains from vegetables whatever is convenient or neceffary for food, drink, cloathing, medicine, navigation, and a thoufand other purpofes.

He has found the means of going down into the abyfs of the earth, and almoft fearching its very bowels. With what artifice has he learned to get fragments from the moft rocky mountains, to make the hardeft ftones fluid like water; to feparate the ufefull metal from the ufelefs drofs, and to turn the fineft fand to fome ufe! In fhort when we follow the feries of created things, and confider how providentially one is made for the fake of another, the matter comes to this, that all things are made for the fake of man; and for this end more efpecially, that he by admiring the works of the Creator fhould extoll his glory, and at once enjoy all thofe things, of which he ftands in need, in order to pafs his life conveniently and pleafantly.

§. 21.

§. 21.

This subject concerning the œconomy of nature, a very small part of which i have lightly touched upon, is of such importance and dignity, that if it were to be properly treated in all its parts, men would find wherewithal to employ almost all the powers of the mind. Nay time itself would fail before even the most acute human sagacity would be able to discover the amazing œconomy, laws, and exquisite structure of the least insect, since as Pliny observes, nature no where appears more herself, than in her most minute works. Every species of created beings deserves to engross one examiner.

If according to gross calculation we reckon in the world 20,000 species of *vegetables*, 3000 of *worms*, 12000 of *insects*, 200 of *amphibious animals*, 2600 of *fishes*, 2000 of *birds*[p], 200 of *quadrupeds*; the whole sum of the species of living creatures will amount to 40000. Out of these our countrey has scarcely 3000,

---

[p] How the author came to reckon 2000 species of birds in the world i cannot guess, for in the Syst. Nat. Linn. edit. 6. there are only about 150 mentioned, and in the last edition of that book not above 550.

for

for we have discovered only about 1200 native plants, and about 1400 species of animals. We of the human race, who were created to praise and adore our Creator, unless we choose to be mere idle spectators, should and in duty ought to be affected with nothing so much as the pious consideration of this glorious palace. Most certainly if we were to improve and polish our minds by the knowledge of these things; we should, besides the great use which would accrue to our œconomy, discover the more excellent œconomy of nature, and more strongly admire it when discovered.

Omnium elementorum alterni recursi sunt,
Quicquid alteri perit in alterum transit.
       Senec. Nat. III. 10.

THE foregoing piece, though on a subject often treated by learned and ingenious men, seems to me to contain many things new and curious, and to give a more comprehensive and distinct view, as it were in a map, of the several parts of nature, their connections and dependencies, than is any where else to be found. But exclusive of this or any other comparative merit, it certainly conveys an usefull lesson,

OF NATURE.

lesson, and such an one as the best of us often want to have inculcated.

From a partial consideration of things, we are very apt to criticise what we ought to admire; to look upon as useless what perhaps we should own to be of infinite advantage to us, did we see a little farther; to be peevish where we ought to give thanks; and at the same time to ridicule those, who employ their time and thoughts in examining what we were, i. e. some of us most assuredly were, created and appointed to study. In short we are too apt to treat the Almighty worse than a rational man would treat a good mechanic; whose works he would either thoroughly examine, or be ashamed to find any fault with them. This is the effect of a partial consideration of nature; but he who has candour of mind and leisure to look farther, will be inclined to cry out:

How wond'rous is this scene! where all is form'd
With number, weight, and measure! all design'd
For some great end! where not alone the plant
Of stately growth; the herb of glorious hue,
Or food-full substance; not the laboring steed,
The herd, and flock that feed us; not the mine
That yields us stores for elegance, and use;

The

The sea that loads our table, and conveys
The wanderer man from clime to clime, with all
Those rolling spheres, that from on high shed down
Their kindly influence; not these alone,
Which strike ev'n eyes incurious, but each moss,
Each shell, each crawling insect holds a rank
Important in the plan of Him, who fram'd
This scale of beings; holds a rank, which lost
Wou'd break the chain, and leave behind a gap
Which nature's self would rue. Almighty Being,
Cause and support of all things, can i view
These objects of my wonder; can i feel
These fine sensations, and not think of thee?
Thou who dost thro' th' eternal round of time;
Dost thro' th' immensity of space exist
Alone, shalt thou alone excluded be
From this thy universe? Shall feeble man
Think it beneath his proud philosophy
To call for thy assistance, and pretend
To frame a world, who cannot frame a clod?—
Not to know thee, is not to know ourselves—
Is to know nothing—nothing worth the care
Of man's exalted spirit—all becomes
Without thy ray divine, one dreary gloom;
Where lurk the monsters of phantastic brains,
Order bereft of thought, uncaus'd effects,

Fate

OF NATURE.

Fate freely acting, and unerring Chance.
WHERE meanless matter to a chaos sinks
Or something lower still, for without thee
It crumbles into atoms void of force,
Void of resistance—it eludes our thought.
WHERE laws eternal to the varying code
Of self love dwindle. Interest, passion, whim
Take place of right, and wrong, the golden chain
Of beings melts away, and the mind's eye
Sees nothing but the present. All beyond
Is visionary guess—is dream—is death.

ON THE

FOLIATION of TREES.

ON THE FOLIATION of TREES;

OR,

The time when they put out their leaves.

By *HARALD BARCK.*

UPSAL, 1753. May 3.

Amæn. Acad. vol. iii.

§. 1.

BOtanists in every age have not only taken great pains to discover and give names to plants, but have also described them with all possible accuracy. But this part of knowledge has been, till this present age, confined to narrower bounds than it deserved; for an opinion has prevailed amongst almost all the men of learning, that it is of no use out of the regions

gions of medicine. From whence it has happened, that we find very few that have cultivated botany, but physicians; nor have even these carried their inquiries farther than to obtain a moderate knowledge of officinal plants. But in our times some, who are worthy of the highest regard from all true lovers of this study, have endeavoured to find out and investigate the vertues of plants with greater care, and industry. For these men besides medical uses have discovered great, and remarkable advantages accruing from such researches.

However i do not intend to give a catalogue of them here, but shall content myself with just touching upon some few things, that have been done in this way, in our own university. In the *Philosophia Botanica* our illustrious president has shewn, that every soil has its own peculiar plants, which we should seek for in vain any where else; and that certain plants keep, as it were, their watches, i. e. expand their flowers and close them again at stated times . The dissertation on the *espousals of plants* has imparted to the learned world the use of various phænomena, which

 Vid. Philos. Botan. p. 263, 273. Barck. This curious subject is amply treated in Amœn. Acad. vol. 4.

occur in the fœcundation of plants. The *Flora œconomica* has faithfully set forth the use of plants in private life. The dissertation on the *buds of plants* has opened to us the cause, why various trees cannot bear the snows, and frosts of our part of the world. From the essay on the *esculent plants* of our countrey we find, that there are many plants growing with us which are proper for food, hitherto overlooked. In the *Swedish Pan*, it is shewn, that certain plants only are destined for sustenance to certain animals. From the *Hospita Insectorum Flora* we are informed that certain vegetables are eat by certain species of insects.

It is now the fourth year since our illustrious president exhorted his countreymen to observe with all care and diligence, at what time every tree expands its buds, and unfolds its leaves; imagining, and not without good reason, that our countrey would some time or other, from observations of this kind made in different places, reap some new, and perhaps unexpected advantage. Upon this admonition, i at that time living in Smoland with that noble person G. A. Witting major, and knight of the military order, was incited to observe for the space of three years, beginning from the year 1750,

1750, the days when different trees began to put out their leaves, when the countreymen sowed their fields, and how much time there passed between seed time, and harvest. This i did with intent, if possible, to find out fixed laws by which to regulate the proper seed-time in every province. But the few observations, which i was able to make, were not sufficient for this purpose; that the work therefore which i meditated might not rest upon too slight a foundation, our president communicated all the papers sent to him from different places for my examination. Such then is the design of this essay, and i submit it to the candid reader, hoping that he will look upon it with an indulgent eye.

§. 2.

Our lands, which lye under a cold sky, are bound up with frost all the winter. Hence the roots of our plants oppressed, as it were, with a drowsy sleep, are benummed, and many herbs, that remain above ground, dye[r]. But when

[r] We have had five winters remarkably severe in Sweden, viz. 1665, $\frac{1683}{1684}$, $\frac{1708}{1709}$, $\frac{1739}{1740}$, and 1751. The cold of which last Feb. 1. N.S. was extremely intense, and such as has

when the sun by its mild rays at the beginning of spring refreshes the earth, the snows melt, the

has scarcely been known in this age, for the botanic thermometer sunk to 32 degrees. Barck.

In that thermometer the freezing point is 0, and that of boiling water 100. So that taking it for granted that the author must mean 32 below 0, this point would answer to 57 below 32 or the freezing point of Farenheit, which is a degree of cold never known in this countrey. I am assured from good authority, that in the year 1739 the thermometer did not sink nine degrees below freezing point in England. They who are curious to see much more surprising instances of cold than that in Sweden, may consult the preface to Gmelin's Flora Sibirica, where they will find how very apt philosophers are to fall into mistakes about the powers of nature, when they trust to theory instead of consulting experience. Monf. Maupertuis says, that the mercury in Reaumur's thermometer in Lapland sunk to 37 degrees below freezing point, which is equal to 67 degrees in Farenheit.

Perhaps, says Linnæus in the Flora Lapponica, the curious reader will wonder how the people in Lapland during the terrible cold, that reigns there in winter, can preserve their lives; since almost all birds, and even some wild beasts, desert it at that time. The Laplander not only in the day, but thro' whole winter nights is obliged to wander about in the woods with his herds of rhen deer. For the rhen deer never come under cover, nor eat any kind of fodder, but a particular kind of *liverwort*. On this account the herdsmen are under a necessity of living continually in the woods, in order to take care of their cattle, lest they should be devoured by wild beasts. The

the ice gives way, the frost is diffolved, and a joyful face of things returns. Immediately we fee

Laplander eafily does without more light, as the fnow reflects the rays that come from the ftars, and as the *aurora borealis* illuminates the air every night with a great variety of figures. The cold is fo great that forreigners are kept aloof, and even deterred from their moft happy woods. No part of our body is more eafily deftroyed by cold than the extremities of the limbs, which are moft remote from the fun of this microcofm, the heart. The kibes that happen to our hands, and feet, fo common in the northern parts of Sweden, prove this. In Lapland you will never fee fuch a thing, altho' were we to judge by the fituation of the countrey we fhould imagine juft the contrary, efpecially as the people wear no ftockings, as we do, not only fingle but double, and triple. The Laplander guards himfelf againft the cold in the following manner. He wears breeches made of rhen deer fkins with the hair on, reaching down to his heels; and fhoes made of the fame materials, the hairy part turned outwards. He puts into his fhoes *flender-eared broad-leaved cyperus grafs*, carex v ficaria, Spec. Pl. that is cut in fummer and dryed. This he firft combs, and rubs in his hands, and then places it in fuch a manner, that it not only covers his feet quite round, but his legs alfo; and being thus guarded, he is quite fecured againft the intenfe cold. With this grafs they ftuff their gloves likewife in order to preferve their hands. As this grafs keeps off the cold in winter, fo in fummer it hinders the feet from fweating, and at the fame time preferves their feet from being annoyed by ftriking againft ftones, &c. for their fhoes are very thin, being made, not of tanned leather, but the raw hide. It was

see the vernal flowers begin to celebrate their nuptials, and the trees, one after another, open their buds, and cloath themselves with leaves. It is a matter of wonder why the *wood plants*, as the *spurge laurel*, the *wood anemone*, the *noble liverwort*, the *vernal vetch*, the *broom rape*, the *pasque flower*, the *colts-foot*, the *sage of Jerusalem, pilewort, violets*, &c. and the garden plants, as the *assara bacca, snow drops, bulbous violet, vernal crocus*, &c. should flower in the very beginning of spring; when we cannot by any pains, or care bring them to flower in the autumn, or after the summer solstice. For it is remarkable that these plants, which are so very patient of the cold in the spring, are yet in the autumn so tender, and weak, that they dye like the Indian plants upon the first hoar frost [1], e. g. the

was difficult for me to find what particular kind of grass they prefer for this purpose, as not being every where the same, tho' always one of the *cyperus grasses*, but I perceived at last that it was what I mentioned above. Thus far Linnæus. I will add, that this grass grows with us.

[1] The iron nights, as they are called in the Swedish language, i. e. sharp nights, happen generally at Upsal between the 19th and 31st of August. e. g. 1746 they began the 19th, 1748 the 17th, 1749 the 1st of Sept. 1750 the 20th of August, 1751 the 27th, 1752 the 20th. They seldom

the *blue mountain thistle, touch-me-not,* &c. On the contrary we see *succories* and *thistles* never flower before the same solstice, whence the husbandman judges from their flowers, as from a calendar that cannot deceive, that the solstice is past. From hence it is evident, that there is something else besides moisture and heat which promotes the fertility of plants.

§. 3.

In the same manner trees observe fixed laws, and a certain order in their leafi g; so that he, who is but moderately versed in this affair,

---

seldom last above three or four nights. After these barley does not grow, and about the time they come on, the gardeners do not venture to trust their green-house and other tender plants any longer to the open air. At that time the leaves of the *fig*, the *mulberry*, the *walnut*, the *vine*, the *toxicodendrum*, and even of the *beech* are shrivelled up. The Indian plants, such as the *kidney bean*, the *African marygold*, the *cucumber* the *amaranth*, the *convolvulus*, the *tobacco*, the *thorn apple*, &c. dye. Nay, sometimes even our native plants, as the *noli me tangere*, the *lesser burdock*, the *bryony*, the *vipers buglos*, the *pimpernel*, the *blue mountain sow-thistle*, the *goosewort*, &c. wither. But before this happens, the *meadow saffron* puts forth its flowers, and that sometimes sooner, sometimes later, according as these iron nights come sooner or later. Darck.

may

OF TREES.

may immediately know, when he sees one species of trees in leaf, what species will be next in leaf. Nor do we hardly ever find this order of Flora transgressed. He who should imagine he had found the true cause of this phænomenon in the different depths of the roots of different trees would be mistaken; for then shrubs would always be in leaf before trees of one, and the same kind; which yet rarely happens. This phænomenon therefore arises without doubt from some other cause, hitherto undiscovered, and perhaps explicable only by the different texture of the tree.

The order of the leafing of trees with us is as follows.

1 *Red elder*
2 *Honey suckle*
3 *Gooseberry*
4 *Red currant*
5 *Spiræa frutex*
6 *Bird cherry*
7 *Spindle tree*
8 *Shrub cinquefoil*
9 *Common elder*
10 *Privet*
11 *Quicken tree*
12 The *osier*
13 *Alder*
14 *Sea buckthorn*
15 *Apple tree*
16 *Cherry tree*
17 *Water elder*
18 *Birch*
19 *Hasel*
20 *Elm*
21 *Dog rose*
22 *Pear tree*
23 *Plum*

23 *Plum tree*
24 *Buckthorn*
25 *Berry-bearing alder*
26 *Lime tree*
27 *Beech*
28 *Aria Theophrasti*
29 *Asp*
30 *Maple*
31 The *oak*
32 The *ash* [t]

With the first soft breeze, says Pliny, the *cornelian cherry* puts forth its buds, next the *bay* a little before the æquinox. The *lime*,

[t] As I do not know that any thing of this kind has ever been published in England, i will subjoyn the order of the leafing of some trees and shrubs, as observed by me in Norfolk, Ann. 1755.

1 Honey suckle	Jan. 15	19 Marsh elder	Apr. 11
2 Gooseberry	March 11	20 Wych elm	12
3 Currant	11	21 Quicken tree	13
4 Elder	11	22 Hornbeam	13
5 Birch	April 1	23 Apple tree	14
6 Weeping willow	1	24 Abele	16
7 Rasberry	3	25 Chesnut	16
8 Bramble	3	26 Willow	17
9 Briar	4	27 Oak	18
10 Plumb	6	28 Lime	18
11 Apricot	6	29 Maple	19
12 Peach	6	30 Walnut	21
13 Filberd	7	31 Plane	21
14 Sallow	7	32 Black poplar	21
15 Alder	7	33 Beech	21
16 Sycomore	9	34 Acacia robinia	21
17 Elm	10	35 Ash	22
18 Quince	10	36 Carolina poplar	22

the

the *maple*, the *poplar*, the *elm*, the *fallow*, the *alder*, the *filberd* and *hasel* are among the first that put out leaves; the *plane tree* also is very early. Nat. Hist. lib. 16. 25.

The foliation or leafing of the first four named trees, 1, 2, 3, 4, varies very much as to the time, and the day on which they break bud; for as the winter goes off sooner or later, so they are in leaf sooner or later. But this does not hold of the rest, e. g. in the year 1750, in which there was scarcely any winter-weather, but the whole was almost a perpetual spring, i observed towards the latter end of March, that the *currant* and *gooseberry* were in blow about Gripenberg; whereas the last year they did not blow till the middle of April. The *oak*, and the *ash* seldom shew their leaves before the night frosts are over [v]. For which reason gardeners do not venture to trust their house plants to the open air, till the leaves of the last trees give sign of a mild winter.

[v] This agrees with lord Bacon's observations, Nat. Hist. p. 146. that a long winter makes the earlier and later flowers come together. This i observed was the case in the year 1755, when the spring was very backward. The author says in a note, that it has been observed for above ten years past, that the oak has been always in leaf before the end of May, in Upland.

§. 4.

§. 4.

The prudent husbandman will above all things watch with the greatest care the proper time for sowing; because this with the Divine assistance produces plenty of provisions, and lays the foundation of the public welfare of the kingdom, and of the private happiness of the people. The ignorant farmer being more tenacious of the ways, and customs of his ancestors, fixes his sowing season generally to a month, and to a day; whether or no the earth be prepared to receive the seed he little cares. From whence it frequently happens, that the fields do not return what might be expected, and that what the sower sowed with sweat, the reaper reaps with sorrow. Wise œconomists therefore in all ages have endeavored to their utmost to fix a certain time for sowing; but hitherto their labor has proved fruitless. There have been some, who have tryed to discover the qualities of the land necessary for this purpose, by taste and smell; nor have there been wanting too others, who were persuaded, that the smell of the earth, and the *fila divæ virginis* *, were infallible signs of seed-time. All

---

* I do not understand the meaning of these words.

which,

OF TREES.

which, although perhaps they are not wholly without foundation, are yet insufficient for obtaining the end we aim at. For the experience of many years has taught us, that the seeds of one and the same species sown in the same ground at different times do not produce equal crops. We have seen even a great difference between what was sown in the morning and the afternoon. Thus also while one plant is vigorous and florishes, another of the same nature, and raised in the same soil withers, and dyes. The farmer often throws the cause of scarcity upon Providence, that means to punish an ungrateful people, by ordering the fields to mourn in weeds, and the corn to mock the threshers toil with empty husks; but it may be with truth asserted, that this surmise is often without foundation. He ought rather to complain of his own imprudence, and accuse himself that his granary is not better stored.

We look up to the stars [u], and without reason suppose that the changes on earth will answer

---

[u] This looking up to the stars for this purpose, was transmitted down to us by the Greeks and Romans from Ægypt, where the seasons being much more regular than in these northern parts, might be as sure a guide in that countrey, as any they could follow. But an astronomical calendar

answer to the heavenly bodies; entirely neglecting the things that grow round about us. We

calendar perhaps may not be so good a guide to us as the vegetation of certain plants; supposing we could once fix on the proper one for sowing each kind of seed. I have been told by a common husbandman in Norfolk, that when the oak catkins begin to shed their seed, it is a proper time to sow barley; and why might not some other tree serve to direct the farmer as to other seeds? The prudent gardener never ventures to put his house plants out, till the mulberry leaf is of a certain growth.

It appears from Geminus in his elements of astronomy, that the coincidence of the seasons, as to heat, cold, rain, &c. with the risings and settings of the stars, had caused a notion to prevail among the antients, that these celestial phænomena were not merely the signs, but the causes of the different seasons. This notion, which he takes some pains to overturn, would never have begun in such uncertain climates, as are found in these parts of the world. But in Ægypt, where the Nile begins to rise regularly upon the appearance of Sirius, or the dog-star, where the Etesian winds begin, and cease to blow constantly about the same time of the year; and in general the variation of the weather is nearly uniform, such a notion might easily prevail in the minds of an unenlightened and superstitious people. From them it was propagated into Greece, where, tho' it must have been frequently thwarted by a much less constant uniformity, yet it might still be upheld by that blind veneration, which generally attends antiquity, especially amongst the ignorant and unlearned. As for the Romans, they went still farther, for without even adapting an almanack to their own climate and time, they fixed the seasons for husbandry work of all kinds by the risings and settings of the stars, such as they

found

OF TREES.

We see trees open their buds, and expand their leaves; from hence we conclude that spring found them in the Greek calendars. To this custom Geminus certainly alludes when he observes, that an almanack, which may pretty well foretell the weather in one countrey is good for nothing in another, as one would think should be obvious at first sight. Yet this he thought necessary to explain, and dilate upon, in order to convince the Romans of their error; for tho', as Petavius observes, the later astronomers went more accurately to work, the prejudice still remained in the minds of the countrey people, and the vulgar. Whether Geminus thought those predictions concerning heat, cold, rain, drought, &c. which are found in the Alexandrian, Greek, and Roman calendars, just as in some of our modern ones, were universally precarious, or whether he only thought they were so in such climates, as that of Rome, where he is supposed to have lived, he commends Aratus for making use of the natural signs, taken from the aspects of the sun, and some of the stars, as also of the signs taken from brutes, instead of the rising and setting of the stars, and gives this reason of his preference, that those predictions which have some natural cause, have a necessary effect; adding, by way of confirmation of his opinion, that Aristotle, Eudoxus, and many other astronomers, made use of them. These predictions are copied by Virgil, but i do not recollect any place in his Georgics, where the seasons for ploughing, sowing, &c. are fixed by the appearance of birds of passage, or of insects, or by the flowering of plants, which method was begun by Hesiod, but never afterwards attended to, that i know, till Linnæus wrote. Hesiod says, that if it should happen to rain three days together when the *cuckow* sings, then late sowing will be as good as early sowing. That when *snails* begin to creep out of their holes, and climb up

spring approaches, and experience supports us in this conclusion; but no body hitherto has been

up the plants, you must leave off digging about vines and take to pruning. That when the *artichoak* begins to blow, and the *grashopper* chirps upon trees, which, as Theophrastus observes, was about the summer solstice, then goats are in full season, &c. That when the *fig leaf* is about as big as a crow's foot, the time for sailing comes on. That when the voice of the *crane* is heard overhead, then is the time for ploughing. It is true, the poet frequently marks the seasons, by the risings and settings of the stars, and as astronomy, besides its many important uses, is connected with finer sciences, has something in it very striking to the imagination, and has been cultivated by men, who had leisure to make calendars for general use, it was natural that it should get the ascendant over rules surer perhaps in themselves, and more adapted to the purpose of the husbandman, but which were destitute of the advantages abovementioned, and were most probably looked on only as poetical embellishments.

It is wonderfull to observe the conformity between vegetation, and the arrival of certain birds of passage. I will give one instance as marked down in a diary kept by me in Norfolk in the year 1755. April the 16th *young figs* appear, the 17th of the same month the *cuckow* sings. Now the word κοκκυξ signifies a *cuckow*, and likewise the *young fig*, and the reason given for it is that in Greece they appeared together. I will just add that the same year i first found the *cuckow flower* in blow the 19th of April.

To the instance of coincidence of the appearance of the *cuckow*, and the fruit of the *fig-tree* in Greece and England, i will been able to shew what kind of tree Providence intended should be our calendar, so that we might know on what day the countreyman ought to sow his grain.

The sun acts on the earth by loosening, warming and preparing it, as the culinary fire does on our meat, for which a certain degree of heat is requisite. For the sun by its heat drives the juices taken in by the roots thro' the vessels of the tree, which do not return by circulation, but become more copious by the daily addition of fresh heat. It. Scan. 23.

i will here add some coincidences of the like nature, in Sweden and England.

Linnæus says, that the *wood-anemone* blows from the arrival of the *swallow*. In my diary for the year 1755, i find the *swallow* appeared April the 6th, and the *wood-anemone* was in blow the 10th of the same month. He says, that the *marsh-marygold* blows when the *cuckow* sings. According to my diary the *marsh-marygold* was in blow April the 7th, and the same day the *cuckow* sung.

I have many other observations by me about the appearance of birds and the flowering of plants, but as they were made for one year only, and there are none of other authors to compare them with, i shall not trouble the reader with them. I have been induced to publish them for reasons that i have mentioned in the preface. Vid. the Calendar of Flora.

§. 5.

Nature always takes the easiest, and shortest way in all her works. He therefore who would imitate her must do the same. No one, I think, can deny but that the same force, which brings forth the leaves of trees, will also make the grain vegetate; and no one can justly assert that a premature sowing will always, and every where accelerate a ripe harvest. Perhaps therefore we cannot promise ourselves a happy success by any means so likely, as by taking our rule for sowing from the leafing of trees. We must for this end observe in what order every tree according to its species, heat of the atmosphere, and quality of the soil, puts forth its leaves. Afterwards comparing together the observations of many years, it will not be difficult from the leafing of trees to define the time, if not certainly, yet probably, when not only *barley*, but *vernal rye*, *oats*, and other annual plants ought to be sown.

§. 6.

To attain this end there were many, who by the exhortation of our president, noted, not only

only the time of the foliation of trees, but the day also on which *barley* was sown, and cut; and were so kind as to communicate to me their observations [w]. I acknowledge myself much obliged to each of these worthy gentlemen for the benevolence shewn me on this occasion, and more particularly to D. Torén, who for the space of three years made his observations on a tree of the same species with care and diligence; as also to D. Eric Ekelund, who did the same with the like industry for two years. Some perhaps had not always time, or opportunity to make their experiments with the same attention; for those, who are detained in cities, often want a number of trees to observe these things as they ought, and those, who live in the countrey, are often drawn by domestic affairs from things of this nature. But if observations were made according to the following rules. 1st, That they should be continued for *three years*, and those specified, as well as the *places* in every observation. 2d, That they should be made on the

[w] The author gives in a note a list of eighteen persons who had communicated their observations made in Sweden, Norway, Finland, and Lapland, some for one, some for two, others for three years from 1750 to 1752 both inclusive.

same *individuals*. And 3d, on trees which grow on the *same soil*, and in the *same exposition*, as the *field* that is to be sown. Were these circumstances, i say, attended to, perhaps we might be able to form more certain rules for the use of the farmer; but since these rules have been sometimes neglected, our business will not succeed so well; for who does not know that the *north wind*, *shade*, and a *moist soil* hinder the leafing of trees as much as a *dry situation* on the slope of a hill inclining to the south promotes it? Besides many errors have crept into these observations, e. g. some trees between whose leafing there ought not to intervene above two or three days, are often disjoined from one another by the interval of a fortnight; not to mention the order of leafing § 3, which trees scarcely, or rather never transgress, being tyed down to it by nature herself, but which often does not appear in these journals[x].

[x] In the original there follows a section which i have not translated. The intent of it is to explain a table giving an account of the different days of the foliation of some trees and shrubs in Sweden, Norway, &c. which i have omitted, as thinking it would afford little, or no entertainment to the reader.

§. 7.

§. 7.

If we confider the year 1750, we may remember, that the winter was milder than ordinary, and the fpring very early. Whence fome in Upland fowed their lands about the end of February; which they fcarcely ever do in other years before April. I am not ignorant, that the lands in fome of the northern provinces, efpecially thofe which abound in clay, require early fowing, that the ground may be broken with lefs trouble, and that the firft fhoots of the barley may make their way thro' it before it grows ftiff. But the people of Schonen, and others, that dwell near the fea, fow late, whether the fpring be early or not; and that fometimes to their great lofs, for no other reafon but that they received this cuftom from their anceftors. The moft northern inhabitants of Sweden find it neceffary to fow as foon as the froft breaks up; that the fhort fummer may perfectly ripen the grain before the winter approaches. For as eggs require a fixed time for the exclufion of the young, fo the barley does in different provinces, to ripen the feed. To prove this i will produce fome examples.

ON THE FOLIATION

		Sowing.	Harvest.	Days.
Pithoa.	1740	June 4	Sept. 1	89
	1741	May 29	Aug. 31	94
	1742	27	29	94
	1743	27	26	91
	1744	31	26	87
	1745	24	27	95
	1746	26	25	91
	1747	28	23	87
	1748	June 4	22	79
	1749	May 21	22	93
	1750	19	14	87
	1751	21	11	92
			Medium	85
Upsal.	1747	April 28	Aug. 17	111
	1748	29	20	113
	1749	May 6	27	113
	1750	April 16	30	155
	1751	28	24	118
	1752	30	31	92
			Medium	105

		Sowing.	Harvest.	Days.
Nasinge	1750	April 20	Aug. 12	113
toward		May 4	7	95
Norway		19	12	85
		21	14	85
		26	15	81
		June 13	25	73
			Medium	93
Korn an	1731	May 28	Aug. 31	95
island of	1732	June 18	Sept. 14	88
Bahus.	1734	May 9	Aug. 18	101
	1735	25	15	82
	1736	29	27	90
	1738	June 3	Sept. 5	94
	1739	May 8	3	118
			Medium	100

From these observations, which i have produced, and many others, i can conclude nothing at present, unless that the *sowing of barley* nearly coincides with the *foliation of the birch*, at least in Upland, and other places adjacent; and if this sign is not to be depended upon every where, yet it would be easy for us, on a due examination, to find out some other tree,

tree, more suited to this purpose; and which some provinces might use as a calendar, while the greatest part might consult the *birch*. It is a popular error, that less time passes between the sowing, and ripening of wheat in our northern provinces, than here at Upsal, and that this happens because the summer days are longer in the north, and there is scarcely any night to retard its growth. But this error is made evident by the grain ripening in as short a time in Schonen as in Lapland. For barley in the champain part of Schonen is sown about May the 29th, and reaped sooner than in Upland. But why *barley* ripens later in Upland and Wessmania, than in the other provinces of Sweden, is to me absolutely a secret.

§. 8.

If a number of future observations shall confirm the doctrine which i have been delivering, i do not doubt but that we may reap many advantages from it. For then we should not want a sure guide for the husbandman to regulate himself by in sowing his grain, and for the gardener to sow his kitchen, and other seeds. What great benefit therefore would arise to the

OF TREES.

public, if one in every province would yearly make obfervations in this way, and at laft communicate them in the fame manner, as aftronomers do their meteorological ones to the royal fociety, or academy of fciences?

It will befides be neceffary to remark what fowing, made on different days in the fpring, produces the beft crop; that comparing thefe with the foliation of different trees, it might appear which is the moft proper time for this purpofe. In like manner it will not be amifs to note at what time certain plants, efpecially the moft remarkable in every province, blow; that it might appear whether the year made a flower or a quicker progrefs. For we fee, although obfervations of this kind have not yet come into ufe, that the mower can guefs at the time proper for cutting grafs, either from the flowers of the *parnaffia*, the *devil's bit*, the *marfh gentian*, or the *baftard afphodel* burfting forth, or from the flowers of the *purple meadow trefoil* withering, or from the ripening of the feeds of the *yellow rattle*, or in higher places from the yellow hue of the leaves of the *leopard's bane*. Would botanifts like aftronomers note the time of foliation, and flowering of trees and herbs, and the days on which the feed is fown, flowers

and

and ripens; and continue thefe obfervations for many years, there can be no doubt, but that we might find fome rule, from which we might conclude at what time grains, and culinary plants, according to the nature of each foil, ought to be fown; nor fhould we be at a lofs to guefs at the approach of winter; nor be ignorant whether we ought to make our autumn-fowing later or earlier. Laftly, the gardener would have a more fure prophet to confult; whereas now he guides himfelf by nothing but very fallacious conjectures.

§. 9.

This is all which i think fit to produce upon this copious fubject, and i hope the candid reader will not be furprifed that i am fo fhort upon it, as it has hitherto not been handled; and is far from being hitherto perfectly underftood. It is much above my power to go to the bottom of this affair, but by touching upon it in a fummary way i mean to excite men of greater ability, who may treat it in the manner it deferves.

OF THE

USE of CURIOSITY.

OF THE
USE of CURIOSITY.

BY

CHRISTOPHER GEDNER.

Upsal, 1752. October 21.

Amæn. Academ. vol. 3.

§. 1.

AS the three kingdoms of nature were created for the use of man, since to him alone is granted the prerogative of converting their inhabitants to his own advantage, so that part of knowledge which is conversant about the creatures throughout the terraqueous globe is the first, and chief by which men are enabled to provide themselves with what is necessary, both for the present and future; and the more so because, besides these three kingdoms, and the elements there is nothing in

nature which can be of use to him. All those things by which man is supported and grows, with which he is cloathed, and in which he prides himself, by which he is preserved, and becomes insolent; all the pomp, the splendor, the richness, the luxury of dress, as well as the necessary covering from hence have their origin. Without these things man must be as naked, as he was created, and came into the world. However obvious this truth may be, there is a common question proposed by the vulgar to men, who are busied in examining the productions of nature, and that with some sort of sneer; *To what end are all these inquiries?* By which they mean to insinuate, that these *vertuosi* are at the bottom but madmen, who spend their time in a kind of knowledge, which promises no advantage; and in this way of thinking they are the more convinced of being right, as they find natural history no part of public institutions, not received into academies amongst the philosophical sciences, and as holding no rank either in church or state. For this reason they look on it as a *mere curiosity*, which only serves as an amusement for the idle and indolent. This objection has been made to myself, and almost all others who give themselves

OF CURIOSITY.

selves up to the study of nature, and by its frequent repetition has at last quite worn out my patience. For which reason i think it will not be amiss to consider the question, and prepare such an answer to those, who for the future shall not be ashamed to urge over and over the same objections, as may convince them, if they will take the pains to read the few following pages, and consider them thoroughly. All i desire of the reader is a candid hearing.

§. 2.

The kind of men, who most frequently ask this question; *To what end all these inquiries?* are of a heavy, dull, and phlegmatic disposition, of weak judgment, and low education. Amongst ourselves, in great cities, in large towns, and at academies, the searching into nature ceases now to be uncommon. Nor is this question ever heard among men of solid learning. It is chiefly, and frequently put in the more remote provinces by the inferior order of people; who think of nothing but indulging their low appetites, and look on every thing as useless, which does not serve that purpose.

When electrical experiments first began to make a noise in the world, Samuel Klingenstierna was sent for by his majesty Frederic the first to shew him some of the electrical phænomena. When all was over, a man of great rank, who happened to be one of the spectators on this occasion, said with a sneer, "Mr. Klingenstierna, of what use is all this?" Klingenstierna replyed with some acuteness; 'Sir, this very objection was made to me by J. C.' (this J. C. was a very rich dry salter). Upon which the king said smiling, to the nobleman, i think he has given it you. Such men as these resemble more the brute creation, than rational creatures. They do not consider, that the all-wise Creator made every thing for man's use. They forget that every thing which was created at the beginning was declared to be good. To these men whatever is curious is disgustfull, and inquiries into nature are deemed mere folly.

Ternstôm (Christ.) when he went with the Ostend fleet to the East Indies, was treated with contempt by some of the company for his curiosity *. They thought nothing of consequence, but what belonged to the winds, and waves.

* Bellonius in his Observations, p. 3. says the same happened to him.

OF CURIOSITY.

Bartfcius (John) when he arrived at Surinam, where he went in order to make obfervations in natural hiftory, was defpifed for looking after plants, and infects. The inhabitants there thought nothing worth minding, but what belonged to *fugar* and *coffé* plantations. Vid. his letters to Linnæus.

Profeffor Kalm dared not at the hazard of his life let the favages of Canada, amongft whom he refided, know that he defcribed any plant or other natural object, but was forced to carry on all his refearches in private.

When our prefident was gathering, and defcribing the *rhen-deer-fly* on the Lapland mountains, the inhabitants wondered, and laughed at him for troubling his head about catching infects. Vid. Act. Stockhol. vol. 1. p. 121. And we find that he, and his companions were ftared at as a fpectacle in his journey through Oeland. It, Oeland. p. 85. 109.

Dr. Haffelquift was forced to have a guard whenever he went out of Cairo in order to defcribe any natural object; and even then he was not quite fafe from the vulgar on account of his curiofity. Thefe examples may fuffice without producing any more.

§. 3.

We were created for the glory of the Creator, which cannot be brought about, unless we know him, either by revelation, or the works of the creation. As to the latter, i suspect, that many come into the world, and remain here even to old age, who never saw the creation, but from afar; just like the brute beasts, which cannot fail of seeing the verdure, and various colors, that cloath the earth, but go not one step farther. This seems to me as if any one, who should be carried into a botanic garden to see the immense variety of plants brought together from all parts with incredible trouble, care and expence, should only observe that the leaves were green, and the flowers of various colors, just as they are every where else. Could such an one be truly, and justly said to have seen the garden? Or if any one should go into a museum, filled with natural objects of the rarest kind preserved in spirits of wine, and should only attend to the clearness of the liquor, and, though he saw a body hanging in it, should not inquire what body it was; would not he, who took the trouble of shewing these sights to so curious a person,

OF CURIOSITY.

person, think his time thrown away? Would such a spectator deserve to be let into such a place?

I cannot help on this occasion calling to mind the manner, in which our president used sometimes to excite attention in his audience by an apt similitude, when he was reading upon *insects* to his pupils. The similitude or rather fable was as follows. 'Once upon a time
'the seven wise men of Greece were met toge-
'ther at Athens, and it was proposed that every
'one of them should mention what he thought
'the greatest wonder in the creation. One of
'them, of higher conceptions than the rest,
'proposed the opinion of some of the astrono-
'mers about the fixed stars, which they believed
'to be so many suns, that had each their pla-
'nets rolling about them, and were stored with
'plants and animals like this earth. Fired with
'this thought they agreed to supplicate Jupiter,
'that he would at least permit them to take a
'journey to the moon, and stay there three days
'in order to see the wonders of that place, and
'give an account of them at their return. Ju-
'piter consented, and ordered them to assemble
'on a high mountain, where there should be a
'cloud ready to convey them to the place they
'desired

' defired to fee. They picked out fome chofen
' companions, who might affift them in defcrib-
' ing, and painting the objects they fhould meet
' with. At length they arrived at the moon,
' and found a palace there well fitted up for
' their reception. The next day, being very
' much fatigued with their journey, they kept
' quiet at home till noon; and being ftill faint
' they refrefhed themfelves with a moft delici-
' ous entertainment, which they relifhed fo well,
' that it overcame their curiofity. This day they
' only faw through the windows that delightfull
' fpot, adorned with the moft beautiful flowers,
' to which the beams of the fun gave an uncom-
' mon luftre, and heard the finging of moft me-
' lodious birds till evening came on. The next
' day they rofe very early in order to begin their
' obfervations; but fome very beautifull young
' ladies of the countrey, coming to make them a
' vifit, advifed them firft to recruit their ftrength
' before they expofed themfelves to the labori-
' ous tafk they were about to undertake.

' The delicate meats, the rich wines, the
' beauty of the damfels prevailed over the re-
' folution of thefe ftrangers. A fine concert of
' mufic is introduced, the young ones begin to
' dance, and all is turned to jollity; fo that this
' whole

OF CURIOSITY.

'whole day was spent in gallantry, till some of
'the neighbouring inhabitants, growing envi-
'ous at their mirth, rushed in with drawn
'swords. The elder part of the company tryed
'to appease the younger, promising the very
'next day they would bring the rioters to
'justice. This they performed, and the third
'day the cause was heard, and what with accu-
'sations, pleadings, exceptions, and the judg-
'ment itself the whole day was taken up, on
'which the term set by Jupiter expired. On
'their return to Greece all the countrey flocked
'in upon them to hear the wonders of the moon
'described, but all they could tell was; for
'that was all they knew; that the ground was
'covered with green, intermixed with flowers,
'and that the birds sung amongst the branches
'of the trees; but what kinds of flowers they
'saw, or what kinds of birds they heard, they
'were totally ignorant. Upon which they were
'treated every where with contempt.' If we apply this fable to men of the present age, we
shall perceive a very just similitude. By these
three days the fable denotes the *three ages*
of man. First *youth*, in which we are too
feeble in every respect to look into the works
of the Creator. All that season is given up to

idle-

idleness, luxury and pastime. 2dly. *manhood*, in which men are employed in settling, marrying, educating children, providing fortunes for them, and raising a family. 3dly. *old age*, in which, after having made their fortunes, they are overwhelmed with lawsuits, and proceedings relating to their estates. Thus it frequently happens that men never consider to what end they were destined, and why they were brought into the world.

§. 4.

As to bodies, the vulgar are ready enough to admire them in the larger kinds of animals, plants, minerals and metals. But when they perceive any one examining into the minute parts of nature, such as *insects* and *shells, grasses*, and *mosses, earthy particles*, and *petrifactions*, they look upon it as idle curiosity. And when they see us searching after such natural productions of forreign countries, as are not found with us, their wonder increases, and they think then they attack us with double advantage. Since we not only spend our time in examining present objects, that are wholly useless, but even such distant ones, as we have scarcely any means of coming at. They
have

OF CURIOSITY. 171

have no notion that these can be of any manner of use but to those amongst whom they are found. To the end therefore that we may gain a clearer conception of the harmony, and use of these things, it will be necessary to run thro' some of the most obvious particulars, relative to this subject, that every one from hence may better comprehend the advantage of natural history in general.

§. 5.

The antients were of opinion, that the bodies about us concerned us no farther than as they were good for *food* or *physic*. Hence their inquiries all tended to find out what were fit to eat, and what would cure some distemper, and whatever plant or animal could not be referred to one of these classes was neglected [y]. It is true that the immediate use of
many

---

[y] I must take the liberty to contradict the ingenious author on this occasion. For any one who has ever looked into Aristotle's history of animals, and Theophrastus's of plants, must at once be convinced of the contrary. This justice I thought due to those two first sketches of natural history, in which the sagacious and extensive genius of the master, and the disciple fully shine forth. It is true this spirit was not long kept up, nor is it to be wondered at, that extravagant speculations, and systems concerning things out
of

many bodies is hitherto unknown to us, yet we have great reason to believe, that all the bodies in the universe, some way or other, contribute to our advantage. *Hay*, which men take such pains to collect in the summer, is of no use to man immediately, but it is a commodity of the utmost consequence to him mediately, as being the food of cattle of all sorts, without which we could not well subsist. Those minute insects called *tree lice*, that live upon the branches of trees, and plants, are looked upon as of no use to us. These are devoured by *flies, cochineals, golden eyes,* &c. in their first state ; which also seem to be of no use to us, but then many of the *small birds* feed upon

of mens reach, which are pursued in the closet with ease, and when ingenious are apt to strike the imaginations of mankind, should take place of the sober, and painful researches into nature, little minded by the generality of people, and therefore lying out of the paths of reputation. Thus what was so well begun by Aristotle and Theophrastus dropped at once for want of encouragement, and never raised its head again, till after the restoration of learning ; when Gesner, Bauhin, Cæsalpinus, &c. in imitation of those first masters, began to revive this part of knowledge; and kindled up a spark which has never been totally extinguished since, and has been raised into a diffusive light by several naturalists of the last age, and particularly by the excellent Linnæus.

them,

OF CURIOSITY.

them, and these not only delight us with their fine songs but afford us most delicate food. The *nettle* is a plant which is scarcely eat by any domestic animal (Iter. Scand. p. 15.) but the Author of nature has allotted to it more feeders than to almost any other plant, v. gr. *butterflies*, *moths*, *wevils*, *chermes*, &c. which devour it almost entirely, and these insects are a prey to many birds, which could by no means live on the plant immediately. *Minute aquatic worms*, and those in no small number, are eat by the larger, and these are eat by the fishes, and aquatic birds, and these by us; and besides food these birds supply us with most delicate soft down to warm and repose ourselves upon. It would be tedious to enumerate all the mediate advantages, which we obtain from the most contemptible; as they are deemed; both plants and animals.

§. 6.

Many look upon *shells* and *corals* of various kinds, which are collected and ranged in museums by the diligent inquirers into nature, as an idle curiosity; since they neither serve for food or physic: but if these are neglected, how many of the wonderfull works of the Creator would

would be unknown? What man of sense is not struck with wonder, when he beholds the innumerable objects, which the author of nature has buried, as it were, in the great abyss. Objects for color, shape, and mechanism so admirable, that they surpass the imagination of man to conceive without seeing them. If we visit a royal palace, and there behold the walls covered with tapestry, pictures, sculpture, and other ornaments, are we not delighted, and even in rapture? We ought therefore to feel the same pleasure, when we behold the beauties of this our globe. To describe every shell on this occasion would far exceed the bounds of my design. At present i will only mention one, viz. the *knotted marginated Cypræa*. Rump. t. 39. f. C. Argenvnill. t. 21. f. K. Petiv. Faz. 97. t. 8. This is a small shell, about the bigness of a hazel nut, and is gathered in the Maldiveè islands by the women along the sea shore in such quantities, that 30 or 40 ships are loaded with them yearly for Africa, Bengal and Siam; so that in those parts there are large palaces filled with them, where they are preserved as treasures of the greatest value.

These shells serve there as gold and silver with us, for all kinds of commerce. In other countries other shells are made use of for various purposes; some instead of horns to blow with at their religious ceremonies; some for vessels for washing; some for cups; some for boxes; some for inlaying; all of them far exceeding the best artificial works.

Nor are those innumerable *petrifactions*, so various in species, and structure, to be looked upon as vain curiosities. We find in our mountains, and even in the middle of stones, as it were embaumed, *animals, shells, corals*, which are not to be found alive in any part of Europe. These alone, were there no other reason, might put us upon looking back into antiquity, and considering the primitive form of the earth, its increase, and metamorphosis. This is a subject, that would require a whole volume to treat it amply as it deserves.

Wild beasts, and *ravenous birds*, though they seem to disturb our private œconomy, are not without their uses, which we should be sensible of, if they were extirpated [z].

When

---

[z] Thus in Suffolk, and in some parts of Norfolk, the farmers find it their interest to encourage the breed of rooks, as the only means to free their grounds from the grub, from which the tree or blind beetle comes. Vid. Lister's Goedact. p. 265. pl. III. Scarabæus. Melolontha. S. N. 10. p. 351. which in its grub state destroys the roots of corn and

When the *little crow* was driven out of Virginia, and that at the expence of several tuns of gold, the inhabitants would willingly have brought them back again at double the price, as we find by professor Kalm. The *vultures* in Cairo are invited yearly, and daily to remain there, as doctor Hasselquist relates in Act. Sac. reg. Scient. Stockhol. 1751. p. 196. et sequ. These creatures of prey cleanse the ground from carcases, and make it wholesome, and pure, and besides they serve to keep up a due proportion between animals, so that one sort may not starve the rest.

The vulgar think, and those who think themselves wiser than the vulgar, make no scruple to say; *let him who has nothing to do employ himself in hunting after mosses and flies.* By which they would insinuate that searching after the minute plants, and animals is unbecoming, or at least unnecessary for a ra-

---

and grass to such a degree, that i myself have seen a piece of pasture land, where you might turn up the turf with your foot.

Mr. Matthews, a very observing and excellent farmer, of Wargrove in Berkshire, told me that the rooks one year, while his men were houghing a turnep field, sat down in part of it, where they were not at work, and that the crop was very fine in that part; whereas in the other part there were no turneps that year.

tional creature. As for *mosses*, i grant we have not authority on our side: for till the end of the last century, they were almost wholly neglected; but now within these fifty years their history is very near compleat by the diligence of Dillenius. C. Bauhin knew very few *mosses*; Dillenius has described near 600. With unwearied pains he went through this very difficult, and extensive branch of natural history. But to what end? it is asked. I will not take upon me to answer this question by shewing the particular use of every *moss* that grows; although i am certain the Lord of nature has made nothing in vain. But i will venture to assert, that posterity will, one time or other, find as many advantages arising from *mosses*, as from other vegetables. I assert this with the greater confidence, because since our acquaintance with *mosses*, we have many experiments, which shew their usefulness, a few instances of which i shall subjoyn. The *bog moss* covers deep bogs with its spongy substance, and thus by degrees turns them into fertile meadows; not to mention its repelling virtue in medicine; at present also its turf is used instead of wood in many provinces, and it is a custom established among the workers in metals to burn it into cinders in their forges. The Laplanders, who lay their children upon it in the cradle, find that it abates the acrimony of the urine. Act. Stock. 1740. p. 421.

The *fontinalis antipyretica*, a kind of *moss*, contrary to the nature of all other *mosses*, guards the walls of houses in case of fire. It. Scand. p. 20.

The *maiden-hair* furnishes a very convenient bed to the Laplander, and the *bear* with this prepares his winter habitation. Most of our *tumps* consist of this kind of *moss*.

The *club-moss* is used for making mats.

The *cypress-moss* furnishes a yellow dye.

The upright *fir-moss* frees cattle from vermin, and purges strongly. It. Oel. p. 28.

The *fountain-moss* points out cool springs.

The *hypnum proliferum*, a kind of *moss*, covers the ground in shady places, where no other plant will grow. Iter. Oeland. p. 28.

The *hypnum parietinum* serves for stopping crevises in walls.

All the kinds of *hypna* and *brya* [a] cover the earth with green, and keep it from being quite naked, as in beech groves and in the woods of both the Indies. They preserve the minute seeds of plants during the winter, shelter their

[a] Names of *mosses*.

roots and keep them from freezing; and gardeners gather *mosses* in the autumn, in order to preserve their plants from the frost; they are gathered by the birds to build their nests; they grow in the most barren soil; by degrees they rot towards the bottom, and thus lay a foundation for fertility.

The *bryum hypnoides* covers the rocks in the coldest mountains.

The *mnium hygrometricum* shews the driness, and moisture of the atmosphere.

Some kinds of *brya* cover the mountains, others the marshes, some are usefull in moist meadow ground, some spread over the naked fields, some are found upon stones, and rocks, others on trunks of trees; and all of them bear the most severe winter, when the generality of other plants grow sickly.

§. 7.

As to the *lichens* or *liverworts*, they are not of less use; for many of them afford a beautifull dye. e. g. the *roccella* yields a most valuable red color, Act. Soc. reg. Scien. 1742. p. 21. to which purpose the *lichen tartareus* serves as a succedaneum. The *lichenes stygius, onuphalodes,* &c. afford also a red dye, and the *lichenes croceus*, *vulpinus* a good yellow. There is no doubt, but that many colors in procefs of time may be obtained from this kind of plants.

If we confider the vertues of the *lichenes* or *liverworts* upon animate bodies taken internally, they are not inconfiderable. The *lichen vulpinus* is a deadly poifon to wolves. It. Scan. p. 40. The *lichen pyxidatus*, or *cup-mofs*, is efficacious in the hooping cough. The *lichen jubatus*, or *rock-hair* in exulcerations of the fkin. The *lichen omphalodes* in ftopping hæmorrhages. The *lichen aphthofus* in thrufhes, and againft worms. The *lichen caninus* or *afh-colored ground liverwort*, in the hydrophia and madnefs. The *lichen pulmonarius*, or *lungwort*, is found to be good in confumptions. The œconomical ufe of the *lichens* is of no fmall confequence. e. g. the *lichen rangiferinus* affords the moft delicious pafture to the rhen-deer. Upon this the whole œconomy of the Laplander turns, and by the help of this, many millions of men are fupported. This *lichen* is alfo given to other cattle by the people of Norland. Act. Soc. reg. 1742. p. 153. Some of the kinds of *lichens* are the delight of *goats*. The moft barren woods, where no other plants grow, afford us

the

the *lichen islandicus*, which in times of scarcity serves instead of bread. Act. Soc. reg. Sc. 1742. p. 154.

The *lichen prunastri*, or *plumb-liverwort*, is ground to powder for the hair.

The *lichen pustulatus* may be converted into a very black pigment. The very *small lichens*, called *leprosus*, cover barren rocks, and makes them look pleasant; it gives birth to black mould, and consequently affords the first degree of vegetative power. After all this can any one justly say that the knowledge of these plants is useless?

The *mushroom* kind also make a class of vegetables by no means to be despised. One species is used in amputations and hæmorrhages, and another is lately come into reputation for stopping the bleeding of arteries; insomuch that the inventor of this use of it, was amply rewarded for the discovery.

The *trufle* and *phalli* contribute to make our soups more delicate, and are commonly used at the tables of the great, Many *mushrooms* are eat by the Muscovites and the inhabitants of other countries, but some of them are a most deadly poison, so that it is of the utmost consequence not to commit mistakes in this part of knowledge.

There is a *mushroom* called *agaricus muscarius*, on account of its driving away flies, and the same plant is the safest remedy hitherto discovered to destroy the *bug*. Thus the knowledge of these plants is of great use to man.

§. 8.

* The *grasses* also are a kind of plants of great value, as affording food for cattle.

The *reed canary grass* serves for thatching houses.

The *meadow fox-tail grass* is an excellent *grass*, which may be sown to advantage in low meadows. It. Oel. p. 156.

The *turfy-hair* causes the meadows in the regio cuprimontana to be so extremely fertile. Act. Soc. R. S. 1742. p. 30.

The *water meadow* is a large and very useful *grass*, which grows by the sides of most ditches and rivers. It. W. Goth. p. 41.

The *narrow-leaved meadow* is the most common pasture in our parts.

The *seed of the flote* or *manna grass*, affords a very pleasing and wholesome nourishment to man.

* Who is curious to know more of these grasses may consult the last piece in the book, intitled, *Observations on grasses*.

OF CURIOSITY.

The *sheep's fescue* makes our sheep very fat.

The *perennial darnel* is the best *grass* for *hay* on chalky hills.

The *sea lyme-grass* and *sea mat-grass* keep the sands on barren maritime tracts from being blown away.

The most minute seeds of *grass* afford nourishment to small birds. The *grasses* besides give a most agreeable color to the earth, and fill up the intervals between plants of other kinds; so that they serve both for pleasure, and utility. The Creator has assigned certain species of *grass* to every different species of soil, which the husbandman is obliged to know in order to make the most advantage of his lands. Besides certain *grasses* are eat by some animals, and left untouched by others; so that without the knowledge of these he cannot avoid falling into error. It. Scand.

§. 9.

He that would exercise the art of husbandry with the greatest advantage, ought to endeavor to get acquainted with all kinds of vegetables, and find out what sort of soil suits each of them best. He ought to know, that some delight in open and exposed situations, others in shady; some in moist ground, others in dry; that some plants thrive most in sandy soils, others in claiey, others in black mould, others in spungy ground, others in watry; some ought to be sown in pools, others on the tops of hills.

Those barren desarts called Alvacu on the mountains of Oeland, It. Oel. p. 206. had long ago been covered with the *crocus*, from whence the inhabitants might have reaped great benefit, if the nature of that plant had been known to them. Our alps, that are more than a hundred miles long, had not remained to this day a mere waste, if our industrious husbandmen, who not long since began to improve the œconomical arts, had known how to cultivate such plants as might have been usefull in food, or physic; and if they had known what usefull trees, and herbs grow on the forreign alps, viz. the Swifs, the Sibirian, the Pyrenean, the Valesian, &c. from whence they ought to have got seed.

The banks of our lakes produce scarcely any thing but *rushes, horsetail, water lilly, pondweeds, reeds*, &c. where nevertheless a great number of plants fit for food might be sown, such as *zizany* of Canada, *water caltrops*, &c.

Every province has its plants, which choak
the

the grain, and render the fields foul, and poor. It. Scand. p. 421. Books of husbandry are full of inventions how to break the earth by instruments, and fit it to receive the seed; this kind of knowledge is insufficient, as long as the husbandman is unacquainted with the nature of those various herbs, to which agriculture ought to be adapted. From hence the necessity of natural history appears.

§. 10.

It is also necessary for the husbandman to know the duration of every plant he sows in his fields, and meadows, viz. whether it be perennial, biennial, or annual. He who wants to know the use of our plants in œconomy, and how few there are, whose use is hitherto discovered, let him look over the *Flora œconomica*. Amæn. Academ. vol. 1[a].

We see how many in a time of dearth suffer for want, fall into diseases, and even perish,

---

[a] The piece here referred to is full of new observations on the uses of plants hitherto not attended to. I wish I could have made such a translation of it, as could have been instructive or entertaining to the public; but a long list of the names of plants, which could have conveyed no ideas to such readers as this work is intended for, must have been very tedious, and very useless.

for no other reason but because they do not know what plants are eatable, and how great a plenty there is of them in our countrey, of which D. Hiorth in this volume has given an account, which the most illustrious senator Baron Lowenheilm has translated into Swedish. Many people wonder why the curious enquirers into nature will give themselves so much trouble about exotic plants; but they do not sufficiently consider, that many kinds of *grain*, many *roots, legumes, fruits, sallads*, and *trees* in common use with us for nourishment, houshold utensils, cloathing, and ornament are originally exotics. Here follows a list of some, which have lately been brought into our countrey from the farthermost parts of Sibiria, that contribute to adorn our gardens, and change our œconomy.

Larkspur, monks-hood, adonis, vetch, cow parsnep, French *honey-suckle, astragalus, othonna, bastard-saffron, greater centory, colombine, dracocephalon, speedwell, claytonica, flax, hyacinth, lilly, lychnis, poppy, cat-mint, yellow-flowered sage, hooded willow herb, hyssop, wild navew, St. John's wort, sow-thistle, saw-wort,* &c. From that distant countrey we have the *robinia's,* and a *honey-suckle,* that make excellent quick-hedges;

from

OF CURIOSITY. 187

from thence we have the Sibirian *nettle*, that serves for making sacks. If we had a more compleat knowledge of plants, that grow in the southern parts of Asia, and America, we should be able to make more ample and usefull experiments.

To preserve our woods we want to be provided with quick-hedges, for which purpose many kinds of trees are serviceable, such as the *gooseberry-bush*, the *black-thorn*, the *white-thorn*, the *barberry*, the *sea buck-thorn*, the *alder*, the *sallow*, &c. provided each be planted in a proper soil.

§. 11.

We have some of our most efficacious medicines, and best spices from the southern parts of the world; and were it not for the curious in botany they had been neglected; as the *lignum colubrinum* was for a long time. What end would it serve to know, that the *senega root* was good against the bite of serpents, unless botanists had also known the plant? And who would ever have dreamed, that our *milk-wort* would answer the same intent? What end would it have served, that professor Kalm was

witness

witness to the efficacy of the Virginia *avens* and the *monaeda* in intermitting fevers, and of the root of the *ceanothus*, and *diervilla* in venereal cases; if we had not learned how to raise these plants? Or to what end would it have served to cross the ocean, and attain the American *water gladiole*, if we had not found out that it was of the genus of our *water gladiole*? The Europeans at vast expence went on buying the *moxa* from China, the *figwort* from Brazil, and the *jachafchapuch* from North America, till it was known that they grew in our own countrey.

§. 12.

There is, as it were, a certain chain of created beings, according to which they seem all to have been formed, and one thing differs so little from some other, that if we fall into the right method we shall scarcely find any limits between them. This no one can so well observe, as he who is acquainted with the greatest number of species. Does not every one perceive that there is a vast difference between a stone and a monkey? but if all the intermediate beings were set to view in order, it would

OF CURIOSITY. 189

would be difficult to find the limits between them. The *polypus* and the *moss* joyn the vegetable, and the animal kingdom together, for the plants called *confervæ* and the animals called *coralline*, are not easy to distinguish, and the *corals* connect the animal, vegetable, and fossil world.

Hence the botanists of this age have been busied about settling natural classes, which is an affair of the greatest importance, and difficulty; but since the vegetables hitherto discovered are not sufficient for that purpose, this part of knowledge is not compleat. It is therefore incumbent on botanists to get acquainted with exotic plants, that they may arrive at the end desired. If all the *columniferous* plants except the *musk-mallow* were known, the *turnera* never could be referred to this order, but that, as soon as it was examined, connected the *turnera* with the *columniferous* plants.

Where the natural classes are settled we find the vegetables so near akin to one another, that we can scarcely distinguish them, as in the *umbellated*, the *siliquose*, the *leguminose*, the *composite*, &c. most of these orders grow in
Europe,

Europe, and therefore could be easily known, and ranged.

He that knows but a few plants gives characters, which are easy to find out, but are insufficient to settle any thing; and therefore tend to confound, rather than to advance knowledge; so that the natural method is the ultimate end of our systematical inquiries. Without this all is a mere chaos, and if the knowledge of vegetables fails, all that use of them is gone, which the learned in this way might discover to the great benefit of mankind.

It is true indeed that vegetables act upon the human body by smell, and taste; but these marks are not sufficient unless we know the natural orders of plants.

These being known, and the vertues of some vegetables being discovered, we may go on safely in the practice of physic, otherwise not. It follows from hence, that he who desires to make any considerable improvement in this branch of knowledge, must endeavor to get acquainted with those plants, whose use he does not know; and thus he is obliged not to neglect the most contemptible. e. g. no body was

OF CURIOSITY.

was able to form a right judgment of the *cascarilla*, who did not know its natural order. No physician would have even suspected, that our *milkwort* would be usefull in the bite of serpents, and inflammatory fevers, unless the principles of botany had led him to it. No one has even thought of trying the *mitreola Americana* against the bite of serpents, which yet, without ever seeing it, we may certainly conclude to be efficacious in those cases, from the *ophiorrhiza Asiatica*, or *true lignum colubrinum* [b]. When botanists knew the abovementioned *turnera*, but were ignorant to what natural class it ought to be referred, no man could

---

[b] This root is known in the East-Indies to be a specific against the poison of that most dreadful animal called the *hooded-serpent*. There is a treatise in Amæn. Acad. vol. 2. upon this subject, wherein the author Joh. And. Darelius undertakes, from the description of such authors as had seen it upon the spot, to ascertain the plant from which the genuine root is taken. It appears in this account that it had puzzled the European physicians, and what had been sold in the shops for it is the root of a very different plant, and of a poisonous nature.

The true root is called *mungos* for the following reason. There is a kind of *weesel* in the East-Indies called *mungutia* by the natives, *mungo* by the Portuguese, and *muncas* by the Dutch. This animal pursues the *hooded-serpent*, as the

could guefs its vertues. But now that we know, that it is of the *columniferous* order, we may without experience be affured that it is of the emollient kind.

Without this knowledge of the natural orders, the *materia medica* would ftill be as uncertain, as amongft the antients, which is of the utmoft importance to us if life and health be fo.

§. 13.

We are ready enough to put a due value on the larger animals, but many look on the minute tribe of infects, rather created to torment, than to be ufefull to mankind. We

---

eat does the *moufe* with us. As foon as this *ferpent* appears the *weefel* attacks him, and if fhe chances to be bit by him, fhe immediately runs to find a certain vegetable; upon eating which fhe returns, and renews the fight. The Indians are of opinion, that this plant is the *mungos*.

That celebrated traveller Kæmpfer, who kept one of thefe *weefels* tame, that eat with him, lived with him, and was his companion, wherever he went, fays he faw one of thefe battles between her and the *ferpent*, but could not certainly find out what root the *weefel* looked out for. But whether the *weefel* firft difcovered this antidote, or not, yet it is certain, adds Darelius, that there is a root, which is an infallible remedy againft the bite of the *hooded-ferpent*. And this he undertakes to afcertain.

grant

OF CURIOSITY.

grant that they are very troublesome to us. But is therefore all care about them to be given up? by no means. On the contrary we ought to contrive means to get rid of them, that they may not destroy both us and our possessions. This cannot be brought about unless we know their nature; when that is known we shall more easily find out remedies against them The use of insects has been sufficiently explained by the noble Carolus de Geer, lord of the bed-chamber to his majesty, in an oration which he made in the academy of sciences at Stockholm. Another of my fellow-students has undertaken to explain what damages insects of various kinds do us, and another now is actually employed in shewing what kind of insects live

---

^e We have lately had a proof that the knowledge of the nature of insects may sometimes be serviceable to us. The sagacious Dr. Wall of Worcester, upon seeing the case of the Norfolk boy, who was cured of worms by taking down a large quantity of white lead, and oyl, guessed that the cure was performed by the oyl, knowing that oyl is fatal to worms and other insects. Upon this he has since tryed oyl in worm-cases with a great appearance of success, an account of which i saw in a letter from him to be communicated to the Royal Society. That oyl is destructive to worms was known to the antients, as appears by Arist. Vid. Hist. Anim. lib. 8. c. 27.

upon

upon every plant [d]. This makes it unnecessary for me to enlarge at present upon the almost incredible mischief insects do us. I will only in a very few words mention, that we shall never be able to guard ourselves against them, but by their means. For as we make use of *dogs*, and other beasts, in hunting down *stags, boars, hares* and other animals, which do us much damage in our fields and meadows; or as *hawks* may be bred up so as to assist us in taking *herons, larks*, and other birds, so also we might make use of the fiercer kinds of insects, in order to get the better of the rest of these troublesome animals.

We shall never be able to drive *bugs* out of our houses, before we introduce other insects that will devour them, v. g. the wild *bugs*, &c.

We have no easier method of destroying *knats* and *flies* which cause us so much disturb-

[d] The two last-mentioned persons hinted at are, I imagine, J. G. Foskahl, and M. Backner, the first of whom has written a treatise shewing the plants which different insects live upon, the last a treatise on the mischiefs done by insects. Both these are published in Amæn. Acad. v. 3.

ance,

ance, than by providing ourselves with the *libellula*, which devours them, as the *kite* does poultry. We oftentimes find our largest trees entirely stripped of their leaves by the *caterpillars* of the *moth* kind, &c. but when we search after them we find they are all eat up by the larger kind of *carabi* called *sycophantæ*; from whence we may learn, that there is no remedy more efficacious in our gardens, where leaves, flowers, and fruits are almost every year destroyed by those *caterpillars*, than gathering and preserving the above-mentioned *carabi* till they lay their eggs, and then placing them at the roots of trees in rotten wood, till they are hatched. And thus we should effectually guard our trees from these inhospitable guests.

§. 14.

But if we do not think it worth our while for any other reason to turn our attention to the works of nature, yet surely for the glory of the great Creator we ought to do it, since in every plant, in every insect we may observe some singular artifice, which is not to be found in any other bodies; and upon comparing these together, we may be convinced, that this does not happen by chance, but was contrived for some certain end, viz. either the propagation, or preservation of the plant or animal with respect to those other bodies. We find how many plants are fenced against the inclemencies of the elements, and the devastations of animals; and how every animal is furnished with some means, by which it may defend itself against the depredations of the rest; so that no species can ever totally perish, which has been created.

Lastly, from the contemplation of nature we may see, that all created things some way or other serve for use; if not immediately, yet by second or third means. Nay we may see, that what we imagine to be most noxious to us is not seldom highly usefull. Without some of these things our œconomy would suffer extremely. Thus were there no *thistles* or *briars*, the earth would be more barren. We ought not to overlook the minutest objects, but examine them with the glass; for we shall then perceive how much art the Creator has bestowed upon them.

He who beholds one of the *jungermannia*, a kind of *wrack*, with a microscope, must be

forced

OF CURIOSITY.

forced to confess, that he beholds a most stupendous, and wonderful phænomenon. Many thousands of people are supported by ryebread, not one of them perhaps ever saw, in how surprising a manner its husks are armed; which any one, who is desirous, may see by the help of a glass.

The day would sooner fail me than matter, were I to take notice of every thing which this subject affords. Let this then be looked upon as the end of created beings; that some may be usefull to man as physic, others as aliment; some in œconomy immediately, others mediately; some vegetables prepare the ground, some protect those which are more tender, others cover the earth with a green, and most beautifull tapestry, and that perennial; some form those groves to which we fly for coolness, others adorn our globe with their most elegant flowers, and regale our nostrils with their most delicious odors. Lastly, all things demonstrate abundantly the omniscience of the wise Creator, who created nothing in vain, but contrived every thing with so much artifice, that human art, however great it may be, cannot imitate the least of his productions. If we neglect therefore to consider these objects,

jects, they would be like pearl thrown before swine. I beseech you then, who ask me with a sneer to what end this or that stone, plant or animal serves, i beseech you to awake, and open your eyes while you live in this world. All these things are not the work of man, but of wisdom itself, which created both thee and me. He has settled an œconomy in this globe, that is truly admirable by means of an infinite number of bodies, and all necessary, which bear some resemblance to one another; so that they are linked together like a chain. For as in our œconomy neither the plough, nor the hedges, nor the dunghill are fit for food, or physic, yet are absolutely necessary; so in the œconomy of nature there are many things that are as necessary, but not immediately. Men reckon their œconomy amongst the chief of human inventions, consider then the sublimity of the divine œconomy. You see therefore that it must at last be granted me according to the opinion of divines and philosophers, that every thing was created for the use of man, and man for the glory of the Creator. Can you then believe, that any thing can be useless that serves not for food, or physic? The Creator has so framed

the

the world, that man should every where behold the miraculous work of his hands, and that the earth should afford an endless variety, seemingly with intent that the novelty of the objects should excite his curiosity, and hinder him from being disgusted by too much uniformity, as it has happened to some wretches, whose station in life placed them above labor, and who wanted curiosity to look into these things. Some objects were made to please the smell, the taste, the sight, the hearing, or other senses, so that nothing can be said to be without its use. That branch of knowledge which serves to discover the characters of natural things, and teaches us to call them by their names, seems perhaps by no means necessary. But let it be considered that the first degree of wisdom is to know things when we see them, i. e. to know them by their names; and without this knowledge scarce any progress can be made. To know the letters of the alphabet, to joyn them into syllables, to understand words is not solid erudition; yet it is absolutely necessary for him who would become learned. Thus the characters and names of things must be thoroughly learned in order to obtain any use from natural history. We find in the journals of travellers, many things mentioned, partly curious, partly usefull, concerning animals, plants, and stones; but those observations can be of no use to us, till we are able to refer each to its genus; that we may make them a part of the system, and know that this curiosity, or use belongs to this or that object, when it happens to come in our way.

§. 15.

If man was created to give praise to his Creator; if the Creator has made himself known to man by creation, and revelation; if all created things are formed with wonderfull mechanism; lastly, if all things were created for the use of man, and nothing but natural things, and the elements can be of use to him; then it may be inquired with the same reason, to what end any other thing was created, as well as man; the supreme Being having created nothing but for a certain end, and for some valuable purpose. We are often ignorant what that purpose is, but it would therefore be impious to say that any thing was created in vain, since he declared that *every thing which he had created was good.* Gen. i. 31.

OBSTACLES

TO THE

Improvement of PHYSIC.

OBSTACLES

TO THE

IMPROVEMENT of PHYSIC,

BY

JOH. GEORG. BEYERSTEIN.

Amæn. Acad. vol. iii.

PREFACE.

Although physic in its whole extent has received great improvements in this age, as most of its parts have been diligently looked into and reformed; yet its chief strength seems to consist in accurate knowledge of diseases, and medicines, and when we turn our eyes on the present times, we find that many simple medicines have been neglected; which so little deserve it, that they rather ought to be revived,

vived, and brought into practice. Which being the case, i have frequently endeavored to find out the cause of this common ignorance. The result of my inquiries i submit to the judgment of the candid reader in this academical exercise, which, though far from compleat, is the best i could produce, and i hope it may prove of some use, and meet with a favorable reception.

Various causes have concurred to bring many medicines into neglect.

1.

Fashion which prevails in physic, as it does in every other earthly thing. Hence physicians prescribe according to certain received forms, not sufficiently considering whether the success answers. To this must be referred the frequent change of remedies.

Brooklime, borrage, buglofs, plantain, saxifrage, are properly only kitchen plants. *Larkspur* is scarcely of any use, but to adulterate syrup of violets, for which purpose it ought not to be used. *Bugle, motherwort, eye bright, polcy-mountain* of Crete, are kept in shops more from custom, than for any good, and sufficient reasons. The *knot-grass* is retained; while

on

IMPROVEMENT OF PHYSIC.

on the other hand the *bear-berry* has been neglected, though an efficacious astringent. The *grass of Parnassus*, and *sun-dew*, have crept into the shop by chance. The *carline thistle*, an excellent remedy in hysteric complaints, is neglected. Those poor wretched plants the *vervains* increase the number of officinals without any merit of their own, and only supported by the testimony of antiquity.

2.

The many theories and hypotheses of physicians that vary in every age. For men have been vain enough to imagine that they knew the immediate causes of diseases, the manner in which medicines operate, and from their principles have undertaken to deduce the vertues of medicines.

Formerly *hot* and *volatile* medicines were used in acute distempers. At present the *acid*, *cooling*, and *diluting* with *bleeding* are recommended. *Musk, ambergris, civet* were looked upon as most efficacious in *eruptive fevers*, now just the contrary. And thus *meadow-sweet, woodruff, musk crane's bill*, may in their turn come into credit, which now are seldom used for driv-

driving out these eruptions; though we may be assured of their vertues by undoubted experiments long since made.

3.

The neglect of specifying distempers. Hence remedies, which are excellent for some diseases in one man; nay even those very remedies that get the name of specifics on account of some very remarkable vertue, when administered to another, are either of no service, or even do mischief; whereas they would perhaps never fail of a good effect, if the species of the distemper were the same. Therefore till physicians regulate the doctrine of diseases in the same manner, that botanists have done that of plants, medicines must be necessarily precarious [*].

Were any one to set about curing the *hæmorrhoidal colic* in a *plethoritic constitution* by *spirituous and hot carminatives,* which are proper for the *flatulent colic* in a *cold,* and *phlegmatic constitution,* he would soon find most fatal proofs of his error. Of this a very remarkable instance may be seen in

---

[*] As well as I remember this observation is taken from Sydenham. But whoever is author of it, most certainly physic must ever be very imperfect, till this grand desideratum be performed.

diff.

diss. med. dni. Arch. Bæck. de medicam. domest.

4.

An hasty and imprudent judgment about poisons, and their difference from medicines which in reality differ only in degrees of strength. Hence our ancestors scarcely ever dared to prescribe the use of plants, which they imagined to be poisonous.

The *laurel* is neither used in consumptions, nor venereal complaints, though an excellent remedy; because it is suspected to be poisonous. The *pasque flower*, whose root is very efficacious in hysteric complaints, is gone out of vogue; because Helvigius knew a person who dyed upon using a syrup made of it; as if all inebriating drinks were to be discarded, because some have lost their senses, and lives by an inordinate use of them. The *lignum colubrinum*[e], that is famous in venomous bites, and the quartan ague, is neglected for the same reason. Scarce any one dares

---

[e] In vol. 2. Amæn. academ. there is a treatise on the *lignum colubrinum*, in which the author undertakes to determine from what plant this root is taken, and observes that druggists, for want of a proper description, have confounded it with two other plants, one of which, and that generally in the shops, is of a poisonous nature.

recommend the use of the *mandragora*, although Schopperus has shewn its vertues in the gout. The *deadly night-shade* is not yet brought into practice, though we have great reason to expect much from it in dispersing tumours of the breast [f].

5.

The abuses of quacks, and their bold, and dan-

---

[f] I cannot omit saying a word or two on the subject of the *deadly night-shade* on this occasion, as the trial of it caused so much noise in this town some time ago. I know the generality of people look on its fate as decided; and that it is destined never to revive again; but that is not clear to me. Some of the faculty still entertain a good opinion of it, and have seen some benefit done by it. *Antimony* was once entirely discarded out of physic, yet we have seen it since become one of the most fashionable remedies in many diseases. New medicines, and particularly of so strong a nature as the *night-shade*, do not come at once into vogue. The not being able to ascertain the proper manner of giving it, the uncertainty in what cases it ought to be used, and how to obviate the inconvenience attending its use, not to mention many other reasons; these, i say, joined together, are fully sufficient to overturn a medicine of the most promising appearance for a time. But whatever may be the fate of the *night-shade* itself, the disinterested zeal of my worthy friend Mr. Gataker to find out some remedy for the most dreadfull and desperate of all diseases; and the candid manner in which every circumstance, relating to that affair, was communicated to the public, must entitle him to the esteem of every humane person.

gerous

IMPROVEMENT OF PHYSIC. 209

gerous experiments. These have made many patients averse to some of the most celebrated medicines, insomuch that a physician dares not prescribe them. For some timid injudicious friend is always at hand to impose upon their weakness, and let them know, that they are going to take a remedy, which had proved fatal to others; not considering that it was owing to the wrong application, and not to the nature of the remedy.

The *hellebore* formerly cured many deplorable distempers, but by the errors of quacks, and their immoderate doses, it has so happened, that it is fallen into disuse; but the *wild cucumber* and *bitter apple* are beginning to revive again. The bark of the *berry-bearing alder* is a very excellent purge, yet physicians have been almost afraid to prescribe it, perhaps terrifyed by the ill success of those daring men above-mentioned, who gave too large doses of it. Many of the moderns for a long while dared not make use of opium even externally.

6.

The timidity, and caution of physicians lest they should hurt their patients by violent remedies.

P   For

For which reason they give rather mild, than efficacious ones, and act the part of spectators, rather than physicians.

For this reason perhaps the disciples of Stahl reject the *bark*; though from ignorance of botany they use the *cascarilla*, which is certainly a very good medicine in shiverings, but not totally void of malignity. Physicians did not for a long while presume to prescribe the *wild cucumber*; which is indeed pretty violent, but by no means so terrible, that it ought not to be used even in the dropsy. For the same reason they did not venture to use the *squill*, whose vertue is very great in thining viscidities; viz. because they did not know the proper dose of either of them. The *gamboge* is neglected, though the Turks have taught us its efficacy in a quartan; and the experiments of our president in the hospital at Stockholm have confirmed their practice.

7.

Small doses of physic. For while physicians have been over-cautious in their prescriptions, they have fallen into the inconvenience of doing the patient no service; and to confess the truth, I suspect they more generally err this way at present; while they order drachms of plants for an infusion, where ounces would be more proper. On the other hand mountebanks, and quacks, men of an intrepid mind, and invincible impudence, oftentimes make a cure, when the physician of probity fails.

If any one were to prescribe only two grains of *rhubarb* for a purge, he might as well do nothing at all. The *honey-suckle* is used in decoctions, but not in the quantity necessary; for which reason its vertue in purifying the blood is known but to few. The dose of the *china root* ought to be large, or no good can be expected from it in venereal cases. Those remedies which are sought for amongst vegetables for curing the venereal disease are perhaps given more sparingly, than they ought.

8.

The ignorance of apothecaries in botany, who often sell one plant for another; by which means, when the desired effect is not obtained, the physician is deterred from the use of them for the future.

For *rad. hermodact*, which is recommended

in the rheumatism, the apothecary sometimes gives the root of the *meadow saffron*; sometimes of one of the *irises*, which differ from it in vertue. Hence the effect of the physician's prescription being uncertain, he is at last obliged to give it up entirely. For the *scabious* they give the *centaury*, Fl. Suec. 708. For the *brankursine*, the *cow parsnep*, 231; the root of the *toothwort*, which is excellent in the tooth-ach, is neglected, because the apothecary does not know, whether it ought to be taken from the *toothwort* 565, or 518, or some other plant. Instead of the root of the *burnet saxifrage*, which is a good astringent in the hæmorrhage, the root of the *burnet* is wrongly substituted. To this may be referred the mistake of selling the *St. John's wort* 624 for the *St. John's wort* 625, which is vulnerary and good in worm cases.

9.

The ignorance of physicians in botany, or their want of care to reject useless, spurious and improper succedaneums.

We suspect that this formerly was the case; but now that the knowledge of botany is carried

carried so far, we have reason to hope, that things will go better. The *acmella* which is very serviceable in the stone, since it is extremely rare, and dear, is to be supplyed out of those plants which are really akin to it. This choice belongs to the botanist: for which reason our president has obliged the world by informing it, that the *sieges beckia*, as nearest allyed to the *acmella*, may be rightly substituted in its room [g]; which Dr. Hasselquist has confirmed by an experiment made here at Upsal upon a young man afflicted with the stone. The skilfull in botany will easily judge that the *German leopard's bane*, as well as the *common*, carries suspicion of poison; yet the former has been looked on as harmless by those, who were ignorant of botany, and the latter dangerous. The *daisy* is cried up in vain on account of the excellent vertue it is supposed to possess. Practitioners, unless they be skilfull in botany, will scarcely allow the *wild rosemary*

[g] Vid. Amænit. Academ. vol. 2. p. 151. where some succedaneums to the Senega root are mentioned, founded on the same principles.

to be a moſt efficacious remedy againſt the hooping cough; which yet is commonly uſed in this diſeaſe by the Weſtrogoths. The *Turkey baum* is kept in our ſhops, altho' much weaker than the *Canarian*, which is excluded. The *white ſaxifrage* and *dropwort*, tho' neither of them has any extraordinary quality, yet hold a place amongſt our officinals. The *mechoacana* is ſeldom uſed, as being of no great ſtrength, yet it is a very proper purge for infants. The *oak of Jeruſalem* is gathered from the *European* plant, whereas both taſte and ſmell inſtruct us, that we ought to get it from the *American*, as a moſt powerful remedy in conſumptions. The plant and ſtalk of *black currants*, no contemptible medicine in the *hydrophobia*, in *feveriſh dyſenteries*, and other contagious diſtempers, are now neglected, as the antients have ſaid nothing about their vertues; which yet are diſcoverable by the ſmell, tho' not by the taſte.

10.

The uſe of compound medicines. Simples are
ſo

IMPROVEMENT OF PHYSIC.

so very rarely used, that the vertues of plants are not known for want of experience.

It is scarcely necessary to produce instances of this assertion. Whoever turns over the writings of the antients will be astonished at the prescriptions, or rather indexes, in which numberless things are mingled together. This affair ought to be looked into, and regulated; that we might not fall under the lash of some future Serenus Sammonicus, who might address himself thus to some physicians:

Ye jumble in one mass such costly juices,
So various in their natures, in their uses;
That the poor patient, who relies upon you,
At once is cheated of his health, and money.

11.

The mixing things together of a different nature. For oftentimes many things are confounded together, which separately administered might assist the patient, and give credit to the physician; whereas mixed they become useless, one destroying the effect of the other.

Thus *watery* mixed with *dry*, *viscous* with *saline*, *glutinous* with *stiptical*, *sweet* with *acrid*,

acrid, *acid* with *bitter*, *sapid* with *nauseous*, mutually weaken each other [h].

12.

The ignorance of the natural classes. From hence it happens that we cannot form any judgment, conformable to botanic principles, of one plant from the knowledge of another. And thus we are afraid of proposing any uncommon plant, being doubtful what we ought to expect from it.

Dogs mercury has been given internally, for want of knowing the natural classes; whereas he, who is qualifyed to reason about the vertues of plants, will allow only the external use of this plant, and in glysters. The *cow parsnep* has been

---

[h] I cannot help applying to this and the foregoing section two verses of that sensible old poet, œconomist, and husbandman Hesiod, tho' in a different sense from what he uses them.

Νηπιοι υκ ισασιν οσῳ πλιον ημισυ παντος,

Ουδ' οσον εν μαλαχῃ τε κ̀ ασφοδελῳ μεγ' ονειαρ.

Which i shall translate for the sake of the unlearned reader. The meaning is as follows. "Foolish man does "not know how much the half is more than the whole, "and what great benefit may be found from the plants "that grow every where about us."

ranked

IMPROVEMENT OF PHYSIC. 217

ranked amongst the *emollients*, although not one among all the *umbelliferous kind* that i know of, is famous for this quality. The people of America ought to give the *mitreola*, Hort. Cliff. for the bite of serpents instead of the *ophiorrhiza*; which if they were to do, they would hardly ever fail of success, if botanists be not greatly mistaken.

13.

The neglect of vulgar medicines easily to be procured. For we owe the very best of our medicines to the vulgar, who have been taught the use of them by necessity, and conceal them as secrets.

We learned the use of the *mezereon* in the cancer from the countrey people. The *noble liverwort* is reckoned a specific in hypochondriac affections by the Gothlanders. The *linnæa* is commonly used by the Ostrobothnians in gouty pains. The common people use *pepper* oftentimes very injudiciously in acute distempers; in eruptive fevers under certain circumstances very rightly. The countrey people taught us the virtues of the *thrush-moss* for sore throats; of the *hop* in in dislocations; and of the *tremella*, Flor. Suec. 1017. for fixed pains in the joynts. They also chew, and blow the fumes of *garlic* into infants to assuage their gripes; or bruise, and apply it to the navel by way of poultice [1].

14.

The neglect of travelling out of Europe. Which would afford us an opportunity of knowing plants, familiar to forreign nations. And I see not why we should be ashamed of learning any thing useful from Barbarians.

It is not long ago that some botanists, who went to America, discovered to us those excellent medicines, the *great water-dock* in the worst scorbutic cases; the *monarda* in intermittents; the *collinsonia* in the colics of lying-in women; the *lobelia*, the *ceanothus*, the *diervilla*, in venereal cases; the *senega root* and *ophiorrhiza* against the bite of serpents and burning fevers. The celebrated Kalm very lately let us know,

[1] Ulluoa observes that some diseases at Carthagena are become fatal, which formerly were not so. Which he attributes to the neglect of the Indian remedies. For he says the old women even now sometimes cure the *chapetonade*, which is one of the distempers he mentions, and formerly never failed to cure it.

that the *water avens* is looked on as a succedaneum to the *bark* by the people of Canada. The *water fig-wort* that corrects *senna*; the *bark*, &c. were communicated by the Barbarians.

15.

The neglect of reading botanical writers, especially those, who in these latter times have faithfully set forth what they knew, by certain experiments concerning the vertues of plants.

Of this kind are Rheede, Sloane, Feuilleé, &c. The use of the *coris* is unknown to most people, who have not seen what Shaw says on that subject. The vertues of the stalks of the *bitter-sweet* purifying the blood were a secret, till our president brought them to light. Before him the apothecaries gave only the *garden night shade*, or the leaves of the *bitter-sweet*, yet few here have found any good effect from them; as we have rarely given this remedy hitherto in sufficient doses. The *rest-harrow* is seldom prescribed, because physicians have not learned its vertues in the Hungarian fever from Scyller. The antients recommended the *cotton-*

ton-thistle in cancerous cafes; but from neglect of reading the ancients, this specific is almost forgot.

16.

Neglect of a method in exhibiting medicines. For instance, physicians expect those vertues from a dryed plant, or in a decoction, which are not to be found but in the fresh plant, or from its expressed juice. Hence it may justly be expected from apothecaries, that they set about cultivating plants; that such, as ought to be used fresh, may be had daily from their gardens.

The *hedge hyssop*, when fresh, purges very smartly and vomits; when old it produces no effect at all. The diuretic vertue of our *water-flag*, which is very considerable, when the plant is fresh, intirely goes off, when it is kept long. Therefore we ought to expect this vertue from the expressed juice, and not from a decoction of it. The *stone crop*, when dry, has none of that efficacy in the scurvy, which is found in it, when fresh. The same may be said of the *house-leek*, the juice of which is celebrated by the Hottentots. The *radish*, the *scurvy-grass*, the *horse radish*, the *garden, water,*

and

IMPROVEMENT OF PHYSIC.

and *Indian cress*, and the *all-sawce*, ought to be sold in the shops fresh, and not dryed; in order to be of any service in the scurvy. The *recent* root of the *rose-wort* is vastly superior to the *dry* in head-achs. Besides it ought carefully to be considered in what part of a plant its vertue resides. Thus it is the *juice* of the *poppy*, that spreads over the brain, as it were, a Lethean drowsiness; and not the *seeds*, for these are eatable. The sagacity of the moderns has reduced the immense number of distilled waters to a very small list.

17.

Neglect in cultivating plants. Hence apothecaries are necessitated to sell plants, which they have had by them many years, and which have lost all their vertues.

The *spikenard* is more durable, perhaps than any other plant; for it will keep its fragrance above an age, as appears by Burserus's Herbary. But other plants are very different in this respect. e. g. the root of *ginseng*, tho' a great restorative, being so very costly, is seldom prescribed; and when it is, it generally has lost its properties thro' age. For which reason we ought to

con-

contrive methods of cultivating it ourselves. Inftead of the leaves of the true *marum*, which has not its equal in art, or nature, the mouldy ftalks of it are generally found in apothecaries fhops. But we would not be underftood as if in all cafes we prefer the cultivated plants to the wild ones. On the contrary the *vipers grafs*, the *goats beard*, the *fuccory* from the fields are fuperior to thofe which the induftry of the gardener has rendered more delicate; on account of the medicinal bitter, which is wanting in their cultivated ftate. See a catalogue of fuch plants as may be raifed with us, in Linn. Mat. Med. p. 212.

18.

The ignorance of phyficians and apothecaries in relation to our own plants. From whence it happens that they are obliged to procure plants from abroad which may be had at home.

Thus our people buy the root of the *rofewort* and root and feeds of the *garden angelica* collected by the Norwegians on our alps, and fold by them to forreigners. For the reft fee a catalogue of fuch plants, as are natives of our countrey, in Mat. Med. above cited, p. 210. If a purge or any other

IMPROVEMENT OF PHYSIC.

other slight medicine is prescribed to a poor countrey fellow, it must be the produce of the Indies, so that they cannot afford to purchase it. Hence people abhor the thoughts of employing a physician or an apothecary.

19.

The ignorance of many forreign plants. Hence we are uncertain whether those which are brought to us be genuine or spurious; and hence also their genera being unknown, we are uncertain about their vertues.

To this head may be referred the *sea lavender*, the *myrobalan*, the *starry anniseed*, the *balsam* of *Copaiva*, the *balsam* of *Peru*, the *gum animæ, caraunæ, elemi,* the *gum rosins* of *myrrh, bdellium, sagapenum,* the *aloes wood, calambac* [k].

20.

The usual custom in apothecaries shops, of providing only drugs of quick sale. Thus they will not procure some whose vertues are now-adays well known, for fear they should lye up-

[k] Hence appears one of the advantages amongst many others that may arise from the voyages of the disciples of Linnæus into the remotest parts of America and Asia, from whence many of our drugs come.

on their hands. It is the bufinefs therefore of the phyfician, who has any regard for his own reputation, and the patient's welfare, to require the apothecary to procure fuch plants, as he thinks may be ufefull.

Simorouba an excellent remedy in the dyfentery, the *fenega root* in venomous bites, the *profluvii cortex* in the diarrhæa, the *camphorata* in the green ficknefs, the *auricularia* in deafnefs, the *Peragua* in the diabetes, the *fouth-fea tea* in the fmallpox, the *ferpentum radix* againft venomous bites, the *wild flax*, a very ufefull purge, are neglected. The juice of the *hypociftis*, and *fungus melitenfis*, altho' powerful medicines in hæmorrhages, and the *herba dyfenterica* [1], which is named fo from its peculiar vertues, have not yet got a place amongft our officinals.

21.

Want of care in gathering fimples at a proper time, and keeping them, when gathered, in a proper manner.

---

[1] I fuppofe the Inula diffenterica L. Conyza Media. R. 174. is here meant, as I find this note upon it in Fl. Suec. edit. 2. 'General Keith told me that the Ruffians, when 'extremely reduced by the bloody flux, in their expedition 'into Perfia, were reftored to health by this plant.'

The

The root of the *avens*, unless gathered in the beginning of the spring, before the sap by nourishing, and pushing out the leaves has wasted its aromatic vertue, will by no means answer what may be justly expected from it. *Rhubarb* ought not to be brought into an apothecary's shop under ten years from the time of its gathering. The flowers of the St. *John's wort* ought to be gathered before they are full blown, that their balsamic virtue may be preserved. The root of the *angelica* is good for nothing unless it be gathered in the winter. *Sloes* ought to be gathered before they are ripe, and the juice pressed out of them in this state, i. e. before the harshness is softened by the frost, if it be designed for an astringent. *Marum* ought to be kept in vessels well closed, lest the volatile part, in which its vertue resides, should evaporate.

———Still an ample field remains,
But not for me, to others i give way,
Who choose a longer course.

AS i do not pretend to underſtand the ſubject of this piece, and therefore cannot ſay how far the obſtacles to the advancement of phyſic charged upon the Swedes ſubſiſt in this countrey, or whether all thoſe obſtacles, which the author has mentioned, be real or not, my ſole motive for tranſlating it was to draw it out of that obſcurity in which it was buried amongſt many other pieces, relating to curioſities of natural hiſtory. I think i may be allowed to ſay a piece is buried in obſcurity, which is only known to a few, who happen to be in the way where ſuch curioſities are talked of; and an attempt to ſpread it over the nation cannot but be right, if the doctrine be ſolid, and affects our practitioners.

Tho' as i ſaid i do not pretend to underſtand the ſubject of this piece; yet i hope the learned reader will excuſe me, if I add one obſtacle more to the foregoing liſt; it is *the notion which has and i believe ſtill does prevail amongſt ſome phyſicians, that the doctrine of ſpecifics is groundleſs, and took its riſe merely from ignorance in natural philoſophy.* I will not undertake to treat this ſubject as the importance of it deſerves; and therefore ſhall

refer

IMPROVEMENT OF PHYSIC.

refer those who choose to look farther into this affair, to a very curious and ingenious book published not many years ago by doctor Martyn, entitled, *Essaies Philosophical and Medical*. The reader may perhaps find there sufficient reasons to incline him to lay some stress on the old-fashioned doctrine concerning the peculiar vertues of some medicines preferably to others seemingly of the same intention. I will add that the phænomena of chemistry give continual proofs of the reality of this doctrine, and afford so many instances of it, that were i so inclined, i could easily fill some pages with them out of Mr. Boyle, and other authors of credit. Ray in his history of plants, p. 49. cites some very curious observations of this tendency from Grew, which are well worth the consideration of physicians. Upon the whole i cannot help thinking that the want of true and genuine philosophy ought rather to be imputed to those who deny, than to those who maintain the doctrine of specifics; and that we might as well undertake to open all locks with one key, as purge all humors with one medicine.

THE
CALENDAR
OF
FLORA,
SWEDISH and ENGLISH.

Made in the YEAR 1755.

Φραζεσθς δ' ευτ' αν φωνην γερανε επακεσης, &c.
Ημος κοκκυξ κοκκυζει, &c.
Ημος δε σκολυμος τ'ανθει κ) ηχετα τετλιξ, &c.
Ημος δη το πρωτον, οσοντ' επιβασα κορωνη.
Hesiod.

ABSOLVENT POSTERI.

TO THE RIGHT HONOURABLE THE Lord Viscount BARRINGTON, SECRETARY AT WAR.

My Lord,

I Embrace with great pleasure the liberty you allow me of dedicating the following pieces to your Lordship. For tho' i must not presume to speak all i feel on this occasion; yet i hope i may without offence, take notice of that most amiable and benevolent disposition, which makes you delight in assisting those, who are incapable of making any return. This is the least that can be said by one, who is himself of that number, and who is desirous to express in a public manner his sincere gratitude and respect. I am,

MY LORD,

YOUR LORDSHIP's MUCH OBLIGED

AND VERY HUMBLE SERVANT,

BENJ. STILLINGFLEET.

PREFACE.

IN my notes on those treatises selected out of the Amænitates Academicæ, which i published not long ago, i marked the day of the month on which certain trees leafed in the year 1755; and likewise mentioned some coincidences of the coming of birds, and the flowering of plants in this and other countries. The instances i there gave were but few, as i could then find no more parallel observations made in other countries to compare mine with. Since that time another volume of the Amæn. Academ. is come out, in which is a small treatise entitled, the Calendar of Flora. This treatise contains an account of the leafing, flowering, &c. of a great number of plants, as also of the departure and return of birds. As these observations happen to be made the very same year in which mine were, and as they are the first of the kind perhaps that ever were made, i was induced to look over my papers again, which i had thrown by as of no consequence; thinking that in these circumstances some use ought to be made of them, as they might prove entertaining at least, if not instructive to those whose genius leads them to curiosities of this kind. I am very sensible how small the number of such persons is, but i am contented to write for those few, nay, indeed i write because there are so few, being willing as far as lies in my power to increase their number.

But it may be asked perhaps by some, even after they have considered all that is said on this subject in the introduction to the following Swedish Calendar, and in the piece De Vernatione

Arbo-

Arborum, why endeavor to increase their number? Are there not idle people enough already? What signifies whether such a plant be in blow or in leaf at the same time with some others; or when such a bird comes or goes, sings or is silent? If we hear the bird sing, and know for what purposes the plant is useful, we know all that is necessary; every thing beyond that is but the wish, or rather dream of enthusiasm, which wants to give an air of importance to its favorite subject. This perhaps may be said by some; but the same way of reasoning applied to other things will shew, that it may possibly be wrong. For instance, the sea swells twice in 24 hours, and the moon passes thro' the meridian circle as often in the same time. Now should it be said, that if we know each of these truths separately it is enough; and that to know farther what relation in point of time one of these phænomena has to the other, is nothing to the purpose; i believe such an assertion would at this time appear absurd, however it might have passed in ignorant ages. I think we may assert universally, that whenever two things, however disparate in their nature, constantly accompany one another, they are both actuated and influenced by the same cause. Now that cause may probably operate on other things that lye within the reach of our powers, and depend on our determination. Thus that constitution of the air, which causes the cuckow to appear about the time, when the fig-tree puts forth its fruit, may indicate the properest season to sow some of our most useful seeds, or do some other work which it imports us to do at a right time; and that time may not be according

cording to certain calendar days, but according to a hitherto unobserved calendar, which varies several weeks in different years. I do not absolutely assert, that we can come to make use of such a calendar, but i desire that others will not assert the contrary at present, but leave this affair to be decided by the only proper way, which certainly must be experience.

We know from Hesiod, that husbandry was in part regulated by the blowing of plants, and the coming or going of birds; and most probably it had been in use long before his time, as astronomy was then in its infancy [*]; but when artificial calendars came into vogue the natural calendar seems to have been totally neglected, for i find no traces of it after his time, whether for good and sufficient reasons i pretend not to determine. That it was laid aside before the time of Aristophanes we have a positive proof in his *Aves*, where he makes Pisthetairus say, 'Formerly the kite 'governed the Grecians, which according to the 'explication of the scholiast means, that formerly 'the appearance of the kite was looked on as a sign 'of spring. He says afterwards, that the cuckow 'formerly governed all Ægypt and Phœnicia, 'because when that bird appeared they judged it 'was time for wheat and barley harvest.'

I shall make no farther mention at present of the use of plants in directing the husbandman, but take this opportunity of making a digression

---

[*] Hesiod himself was one of the earliest of the Greek astronomers. He lived, according to Sir Isaac Newton, about 70 years after Chiron, who formed the constellations for the use of the Argonauts; and from Hesiod the gross and coarse method of astronomy was called the Hesiodean method.

about

PREFACE.

about birds in relation to their prognostic nature. Henceforward then, i. e. from the time of Hesiod, they seem to have been looked upon as no longer capable of directing the husbandman in his rural affairs, but they did not however lose their influence and dignity; nay, on the contrary, they seem to have gained daily a more than ordinary, and even wonderfull authority, till at last no affair of consequence, either of private or public concern, was undertaken without consulting them. They were looked upon as the interpreters of the gods, and those who were qualified to understand their oracles were held among the chief men in the Greek and Roman states, and became the assessors of kings, and even of Jupiter himself*. However absurd such an institution as a college of augurs may appear in our eyes, yet like all other extravagant institutions, it had in part its origin from nature. When men considered the wonderful migration of birds, how they disappeared at once, and appeared again at stated times, and could give no guess where they went, it was almost natural to suppose, that they retired somewhere out of the sphere of this earth, and perhaps approached the ætherial regions, where they might converse with the gods, and thence be enabled to predict events. This i say was almost natural for a superstitious people to imagine, at least to believe;

---

* Jovi optimo maximo se consiliarum atque administrum datum meminerit augur. Cicero.

Lacedæmonii reges augurem assessorem habuerunt. Id.

Aves internunciæ Jovis. Id.

Sacerdotum collegium vel nomine solenne. Plin. Nat. Hist. speaking of the augurs.

PREFACE.

as soon as some impostor was impudent enough to assert it. Add to this, that the disposition in some birds to imitate the human voice must contribute much to the confirmation of such a doctrine. This institution of augury seems to have been much more ancient than that of aruspicy; for we find many instances of the former in Homer, but not a single one of the latter that i know of; though frequent mention is made of sacrifices in that author. From the whole of what i have observed, i should be apt to think that natural augury gave rise to religious augury, and this to aruspicy, as the mind of man makes a very easy transition from a little truth to a great deal of error.

A passage in Aristophanes gave me the hint for what i have been saying. In the comedy of the *Birds* he makes one of them say thus: 'The
' greatest blessings which can happen to you
' mortals are derived from us; first we shew you
' the seasons, viz. spring, winter, autumn. The
' *crane* points out the time for sowing, when she
' flies with her warning notes into Ægypt; she
' bids the sailor hang up his rudder and take his
' rest, and every prudent man provide himself
' with winter garments. Next the *kite* appear-
' ing proclaims another season, viz. when it is
' time to shear your sheep. After that the *swal-*
' *low* informs you when it is time to put on sum-
' mer cloaths. We are to you, adds the chorus,
' Ammon, Dodona, Apollo; for after consult-
' ing us you undertake everything; merchandize,
' purchases, marriages, &c. Are we not then to
' you on the footing of Apollo, &c.' Now it seems not improbable, that the same transition

was

was made in the speculations of men, which appears in the poet's words, and that they were easily induced to think, that the surprising foresight of birds, as to the time of migration, indicated something of a divine nature in them; which opinion Virgil, as an Epicurean, thinks fit to enter his protest against; when he says,

Haud equidem credo quia sit divinitus illis Ingenium.

But to return to Aristophanes. The first part of the chorus from whence the afore-cited passage is taken, seems with all its wildness to contain the fabulous cant, which the augurs made use of in order to account for their impudent impositions on mankind. It sets out with a cosmogony, and says, that in the beginning were Chaos, and Night, and Erebus, and Tartarus. That there was neither water, nor air, nor sky; that Night laid an egg, from whence, after a time, Love arose. That Love, in conjunction with Erebus, produced the bird kind, and that they were the first of the immortal race, &c.

With this passage in Aristophanes, the account of the oracle of Dodona seems to agree. This oracle was the oldest in Greece, and there a *dove* prophesied, according to the concurrent testimony of history; but according to the explication of Herodotus this strange opinion arose from hence, that the Theban priestess, who was stolen by the Phœnicians, and carried into Greece, was called a *dove*, because being a barbarian, she seemed to the Dodoneans to chatter like a bird, till she had learned the Greek language, and then she was said to speak with a human voice. This
explication

explication seems to me extremely forced, and every thing is much better accounted for by supposing, that at Dodona natural augury was first changed into religious augury; for there the oaks also prophesied; which plainly shews the first state of religious augury, when it had not wholly put off its antient form, but like the monsters in Ovid's Metamorphoses, still retained enough of it to convince us what it had once been. That Dodona was one of the first places where augury was practised, is highly probable; for it is mentioned by Homer as an oracle of established reputation at the time of the Trojan war: now Pliny tells us, that Tiresias invented augury and aruspicy; and that he was reputed an augur appears by Sophocles in the Œdipus Tyrannus, where he is introduced saying thus to Tiresias, 'If you have received any information concern- 'ing the death of Laius from the birds, or by 'other means, do not envy it us.' Tiresias therefore, according to Sophocles, lived in the time of Laius; and Laius, according to Sir Isaac Newton, lived not 80 years before the taking of Troy.

I will here subjoin an account of what has been observed about the disappearance of birds, which will serve to confirm what I said above concerning the effect, which that phænomenon might not improbably have on the minds of men; and give room for the superstitious impostures that arose from thence. Aristotle has a chapter on that subject; wherein he says, 'that many birds, 'and not a few, as some imagine, hide themselves 'in holes;' he then enumerates the *swallow*, the *kite*, the *thrush*, the *starling*, the *owl*, the *crane*, the *turtle*, the *blackbird*, and the *lark*, as certainly hiding themselves; which shews how little was known of their real state in his days; nay, so much was he puzzled about this subject, that in another place he supposes some of the birds to be changed in their form and voice at different seasons. Thus he says, that the *redstart* changes into the *robin redbreast*; and Gesner gives this reason for Aristotle's falling into this opinion, that during the summer the *robin redbreast* lives in desert places, and comes towards towns and houses in the autumn, when the *redstart* disappears. Again Aristotle says, that the *black cap* changes into a *beccafigo*, which last appears, as Gesner observes, about autumn, when the figs are ripe, and the former after the autumn. It is true Aristotle mentions some kinds of birds which go to warmer climates when they disappear, which is a proof that their migrations were not wholly unknown in those days; and indeed the poems of Homer prove that they were in part known much earlier. Nor could it happen otherwise, when the inquisitive genius of Greece began to work, and carry men into Phœnicia and Ægypt, with a view of improving themselves in all parts of learning; where they could not avoid observing, that some birds which left Greece in the winter were found at that time in those warmer climates. But the superstition was already confirmed before this happened. Dodona was established on a foundation not to be shaken by the weak attempt of reason and experience. The birds had given good advice time out of mind, and brought many a general and a magistrate, as well as private men without number,

out

PREFACE.

out of difficulties; and therefore, whether they wintered in Ægypt or not, signified little; and indeed it was only supposing them to go a little further, viz. into Æthiopia, and there they might meet Jupiter at his annual visit, μεθ' αμυμονας Αιθιοπηας, and have the gift of prophecy conferred upon them, or confirmed. Agreeably to these notions we find several birds were looked upon as sacred to particular gods; thus the owl to Minerva, the peacock to Juno, the eagle to Jupiter, the crow to Apollo whose messenger he was called as appears by Hesiod.

Some will be apt to think that i have dwelt much longer upon this subject than it deserved; but i cannot help thinking, that even the infirmities of the human mind, especially such as have like this prevailed amongst the most ingenious and sagacious people we read of, and for a long course of time influenced their most serious concerns, ought to be looked upon as not below our notice.

It may seem wonderful to some, that naturalists have been so long without being able to determine any thing certain about the state of several birds when they disappear. The best writers have given it as their opinion, that *swallows* lye under water all winter; one of the latest ornithologists, a writer of great character, falls into this opinion, and the author of the following Calendar adopts it; and indeed till Monf. Adanson cleared up this point, it must appear a problematical point to any man. But though the migration of this bird is at last determined, yet what becomes of the *nightingale*, the *cuckow*, the *goatsucker*, and several others, is still undecided.

Nor is this wonderful, though it may seem so; for the generality of mankind, and especially those who travel merely for the sake of a livelihood or a fortune, are so little solicitous about things of this kind, that the air might be filled with *swallows* in winter without their observing it, as was plainly the case at Senegal.

The number of birds that disappear in this kingdom is much greater than is generally imagined; especially if we reckon amongst them the birds which shift quarters at different seasons, but do not cross the seas. I shall not attempt to give a list of them, but recommend it to the curious, who live in the countrey the year round to watch them more narrowly than they have hitherto been. Linnæus says, that most of that genus of birds, which he calls *motacillæ*, i. e. *those small birds, which have a beak subulated and strait, with chaps nearly equal, nostrils of a pointed oval form, and tongue jaggedly indented*, live upon insects and not grain; and therefore migrate from the northern to the southern parts towards winter; but it appears, that many birds migrate not only in Sweden, but in Greece and other climates, that live with us all the year round.

It is possible, that after all i have said, tending to revive natural augury, and after all the necessary observations shall have been registred, that no use can be made of it; but i am certain, that as long as men have ears and eyes, they must think that one of the greatest delights of the countrey, especially during the spring months, is owing to the lively motions, beautiful shapes and colours and melodious notes of birds, which will

PREFACE.

will afford more pleasure as they are more observed; and therefore, i am not surprised, that Peter the Great of Muscovy, did not think it beneath his attention to endeavour to enliven his new seat of empire, by sending for colonies of them from other parts, as they were scarce where he resided.

I will finish this digression with a reflection that occurs to me on the different fates of natural and religious augury. The first was simple, unattended with any of those circumstances that are apt to rouse the passions of man; and therefore, tho' likely to prove useful, if pursued with proper diligence, fell into neglect. The latter was complicated, applying itself to some of the strongest passions in man, and therefore, though unlikely to a serious mind, to have the least foundation in truth, or ever to be useful, was encouraged and adorned with all the pomp that a superstitious people could invent in honor of a flattering, and therefore favorite art.

I shall now come to some points that more immediately relate to the following Calendars.

1. I have retained the Linnæan names of every plant, and animal in the Swedish Calendar; and have added the English names to the plants taken from Ray's Synopsis, and his history, with no small trouble, as any one will easily believe who has done the like*. The numbers which follow the English names refer to the abovementioned books with an H. to distinguish the

---

* This trouble we shall for the future be relieved from, when that accurate and skilful botanist, Mr. Hudson, has published his Flora Anglica, which is now in the press.

PREFACE.

history. The numbers after the English names of animals refer either to his Historia Avium, or Piscium, according to the subject §. I chose to refer to Ray, as well as barely give the English names, for the ease and satisfaction of such as put a due value on that inestimable writer, whose works do honor to our nation, as a late disciple of the great Swedish naturalist justly observes. I cannot help saying farther upon this occasion, that no writer till his time ever advanced all the branches of natural history so much as that sagacious, accurate, and diligent English observer, whose systematical spirit threw a light on every thing he undertook, and contributed not a little to those great and wonderful improvements, which have been since introduced.

2. I have omitted most of the plants which are not natives of England, both because it is not easy to find English names for them which have any authority, and because i had scarcely any observations in my own Calendar, but on such plants as are native. Some foreign ones however i have retained, particularly such as are common in almost every garden; and such as are marked in the Calendar, as more than ordinary prognostic. These last are printed in large characters.

3. I have retained the division of months according to budding, leafing, flowering, &c, tho' i could not imitate this method in my own Calendar for want of more experience; but i am

§ Some perhaps may think that i need not have referred to Ray for birds so well known as several mentioned in the Calendars; but the want of this caution in many authors, has produced great confusion and doubt about the things meant in every branch of natural history.

convinced

PREFACE.

convinced that this method marks more precisely when we may expect the flowering of any plant, or the return of any bird, &c. than the bare mention of the day of a common calendar month, and at the same time marks it more universally. Thus when Aristotle says [*], That the nightingale sings continually day and night for fifteen days about the time when the young leaves begin to expand and thicken the woods, he not only marks the time when they might expect to hear the nightingale in Greece, but in every other countrey; for thus it happens in Sweden and England, as may be seen in the following Calendars; whereas if he had said, it appeared in such a day of the month, it would bear true perhaps for that year only; and in fact we find in the old almanacks the same author marking days very distant from one another, for the appearance of the same birds, and thus it must be likewise in relation to plants.

Thus far for the Swedish Calendar. As to my own, 1st. i have marked every circumstance down as i found it in my journal, and hope the learned reader will pardon any mistakes which might happen, either from want of judgment or attention. It is possible that i might put down some plants as first being in bud, or flower, or

---

[*] His words are ὅταν τὸ ὄρος ἤδη δασύνηται. i. e. when the mountain is thickening, where it is certain the word *mountain* is used for the trees which generally grow upon it. Thus Homer applies the word σκιόεντα to ὄρεα for that reason. Iliad. Λ. 157. and Eustathius upon the place says, ἰστεον δε ὅτι, σκιόεντα μεν, ἀπο του παρακολουθεντος, λεγεται δασεα κ᾽ σκιας αποτελεστικα δια το της ὑλης δασιον.

Pliny translates this passage, *densante se frondium germine.*

leaf, because i happened then first to observe them, or they might be in those states some time before in some place where i happened not to go.

2. I wanted such a guide as the ingenious author of the Swedish Calendar. My observations then perhaps might have been less unworthy of the public, as they would have been better directed to a particular purpose; but now the reader must expect to find in it all the imperfections that generally accompany first attempts of any kind.

3. I have caused all the prognostic plants, which are mentioned in my Calendar, to be printed in large letters as in the Swedish. The other marks i shall explain in a page by itself, for the more easy recurring to it.

4. These two Calendars would perhaps upon comparison have furnished me with some observations, had i been able to find time sufficient for that purpose; but a strong desire to communicate them to the public early in the year, that others might be induced to keep journals of the same kind, determined me to send them out in this naked condition; and the more so, as i am assured on very good authority, that such journals will be kept in Sweden, Germany, Italy, and France, the next year; and i think it would be pity, that an opportunity should be lost of making so curious a comparison between these different climates, and which perhaps may not occur again, or at least not for many years.

5. The observations on heat and cold were made with a thermometer, marked in a way peculiar to myself. The degrees are those of Farenheit, which i chose as being in common use,

but

PREFACE.

out instead of 32, i have made o the freezing point. This method is more simple, natural, and uniform, and conveys a more distinct idea to the mind. To this scale i have reduced the Swedish author's observations, as well as those of Dr. Hales, taken from his Vegetable Statics; who i am pleased to find has made use of the method above-mentioned, in his late works, and i wonder it is not universally adopted. The degrees below o i have marked thus,—1.—2.—3, &c.

6. My botanical observations were made on plants growing in the fields chiefly; the Swedish plants growing in the Upsal garden; which method is best, where either is in our power, i cannot determine. There are conveniencies and inconveniencies attending each; but there is one great convenience visibly on the side of the garden; which is, that the plants lye within a small compass, and therefore may be looked over more surely and regularly every day.

7. I once designed to place the two Calendars over-against one another, in opposite pages, part by part, according to the days of the month, but upon consideration i found, the climates being so different, that there would be great vacancies in many of the pages; at the same time that the same plants would be in different pages, and the bulk of the book would be increased without any advantage to the reader; i therefore thought it would be better to make an index, which will furnish an easy method to the curious of comparing the two climates.

8. If ever any use be made of Calendars of this kind, it must be by finding out, after a long series of observations, and publishing by itself a

list of a few regularly prognostic plants, either common in every field, if native; or, if not native, common in every garden. For it must be noted, that many plants will blow even in the depth of winter, if the weather be mild. This is the case of *dandelion, chickweed, shepherd's purse, daisy*, &c. As for other precautions, I will refer the reader to the piece concerning the leafing of trees in the Amæn. Academ.

This Calendar was made at the hospitable seat of my very worthy and ingenious friend, Mr. Marsham, who has likewise made observations of this kind, and lately communicated to the world his curious observations on the growth of trees. All the countrey about is a dead flat; on one side is a barren black heath, on the other a light sandy loam; partly tilled, partly pasture land sheltered with very fine groves.

THE
CALENDAR of FLORA.

By ALEXAND. MAL. BERGER,

Upsal 1755.   Latitude 59. 51.¼

Poma dat autumnus, formosa est messibus æstas
Ver præbet flores.   OVID.

INTRODUCTION.

BEFORE i set forth the Calendar of Flora, or the delights of the year, arising from mere sublunary things according to its progress, and that from observations made in the climate of Upsal, ann. 1755, i think it necessary to say something by way of introduction. Time moves on slowly; every thing is in progression and motion, and has its allotted time, as the wisest of men Solomon observes; to which purpose Virgil says, *Stat sua cuique dies.*

Astronomers have exerted all their power to measure time. To them we owe the accurate divisions of it; for they by observing the course and motion of the celestial bodies, have been at last enabled to reduce it to stated periods, and to divide it in such a manner into years, months, weeks and days, that we have calendars constructed for common use, as a rule by which to observe and number its equal parts.

As the stars radiate, shine, and adorn the celestial regions of the summer months, so flowers beautify and illuminate the earth with a wonderful variety of bright and delightful colors. Thus, according to the stile of the chymists, that which is above is as that which is below.

How much time soever and labor botanists have bestowed for many ages back, in order to know the names, nature and vertues of plants, they have not hitherto arrived at that degree of perfection, as to be able to equal the success of astronomers, in noting the properties and phænomena of each of them.

Every

INTRODUCTION.

Every flower has its appointed season. It would therefore be in vain for us to seek the spring plants in autumn, and the autumn plants in the spring. We see them at stated times emerging, stalking, flowering, fruiting, decaying. Again in another season we see others rising in their room, and that in so short a time, by so regular and constant a law, according to the direction of their natures, that it seems impossible for any one to behold this series and variety, without the highest admiration.

The sun at the same time that it raises, as it were, to life these beings, that are destitute of animal functions, brings them forth also sooner or later, according to the nature and disposition of each, i. e. as this or that plant requires a lesser or a greater degree of heat, before it can obtain its just maturity. For as eggs, differing in species, when sit on by a hen, will not all be hatched the same day, but some sooner, some later, so neither do flowers come forth together, but at stated times, as they shall have received the degree of heat proper to their natures.

Altho' the year was formed by the Creator in such a manner, as to be divided into distinct parts, by the sun sending forth its rays equally on the surface of the earth, yet we are not therefore from thence to define and measure the summer, the quantity of ice and snow and Northern colds hindering the air from being equally soon warmed in different years; and according to the heat of the air, the seasons are advanced or retarded, and this is best known and measured by the various kinds of flowers.

Since therefore the summer season depends
upon

INTRODUCTION.

upon the greater or lesser degree of heat, so that flowers come forth proportionably to those different degrees, but yet in such a manner, that one species follows another in a regular order; since this is the case, i say, the seasons of the year, and particularly the summer, may easily from thence be measured; which hitherto has been a desideratum, on account of œconomical uses, in spite of all the assistance from astronomers.

Hence plants in different years often flower a month sooner or later, although, as i observed before, they still follow one another in their natural order, as far as the summer solstice; at which time they hardly ever differ in any year; and in the same manner they proceed, hasten on, or are retarded, the nearer they approach towards autumn, and the winter is farther off, or nearer at hand. Now in order to determine accurately the acceleration or retardation of the winter, we must observe all the different kinds of flowers in every place, at what time they first appear, and this daily, that the order which they observe may be better ascertained.

By way of specimen i have exhibited the flowers in the same order in which they appeared the last year, 1755, in the Upsal garden. I must observe, that almost all the plants mentioned in the following calendar grew in the open air, and in the same kind of soil, which is low and loamy, excepting about half a score, which were gathered in the woods not far distant, and which are with difficulty raised in the garden.

I have marked the month and day all along
on

on the side of the page, not meaning that any one should thence imagine that the flowers will return every year on the same day and month, but with intent to set forth the calendar of that individual year, and that it might appear with what diligence and circumspection it was made.

In order to distinguish the cultivated plants from the wild, i have used italics for the first, and have marked the plants which appear to be most prognostic by an asterisk *.

I have besides thought fit to dispose them all into months, according to a division the aptest i could contrive; but i did not think it necessary to form equal months, as my design was not to determine days, but chiefly and indeed only the greater or lesser acceleration of summer.

Having so accurately observed the flowers, i thought fit to add the time at which the animal kingdom undergoes certain alterations, such as when birds of passage come and go, when birds of every kind lay, or hatch, or moult, when several kinds of fishes celebrate their nuptials near the sea shore, when it is sowing time, when grain flowers, ripens, &c.

By the help of such observations we may at last come to know what is to be done, or observed, every day, by the flowering of plants. But much time is required to bring this to perfection; and he who observes, must, if he means to do any thing to the purpose, live in the countrey, where it is much easier to see every thing that presents itself.

* Instead of an asterisk i have used great letters.

INTRODUCTION.

If the gentlemen of our own or other countries, took delight in such observations, they might amuse themselves very agreeably, by giving up some of their time to things of this kind; and i am most certainly persuaded, that this so slight a sketch, gaining continually new additions, would at last produce a work of great use; as it might furnish materials for directing private œconomy, and the more so as the times for sowing of seeds, for reaping, and mowing, and for gathering fruits of various kinds, might from thence be best settled.

Gardeners might thence learn at what time of the spring they ought to lay the roots of plants bare, when to sow their seeds, when to expose to the open air, and when to put under shelter their tender plants, and how to furnish the garden with flowering plants; so that there might be a perpetual blow all possible months of the year; thus the *lilac* follows the *cherry*, the *mock orange* follows the *lilac*, and the *late roses* follow the *mock orange*.

THE ORDER OF BLOWING OF THE BULBOUS PLANTS IN BORDERS, AND THEIR DURATION. N. B. The plants are numbered from the first day of budding, by the figures bered on the left hand, the other figures on the right hand shew the duration of their blow*.

---

* The meaning is this, as explained to me by Mr. Solander; suppose the *snow-drop* buds on any given day, then the *crocus* will bud the second day after it, the *hyacinth* the twelfth day after it, &c.

1. Snow-

INTRODUCTION.

1. Snow-drops, 1144. H. Galanthus *nivalis*, 26.
Violet, *bulbous*, 1144. H. Leucoium *vernale*, 26.
2. Crocus, *spring*, 1174. H. Crocus *vernus*, 17.
12. Hyacinth, *oriental*, 1159. H. Hyacinthus *orientalis*, 18.
20. Fumitory, *bulbous*, 975. H. 4. Fumaria *bulbosa solida*, 20.
23. Hollow-root, 975. H. 5. Fumaria *bulbosa cava*, 20.
28 Hyacinth, *grape*, 1161. 28. II. Hyacinthus *botryoides*, 19.
34. Daffodil, *English*, 371. 2. Narcissus *pseudo-narcissus*, 19.
Daffodil, *sweet*, Narcissus *odorus*, S. N. 19.
37. Crown, imperial, 1105. H. Fritillaria *imperialis*, 10.
Fritillary, *Pyrenean*, 1107. H. Fritillaria *Pyrenaica*, 14.
38. Lilly, *chequer'd*, 1106. II. Fritillaria *meleagris*, 10.
44. Tulip, 1146. II. Tulipa *Gesneriana*, 13.
47. Primrose, *peerless*, 1133. H. Narcissus *poeticus*.
50. Hyacinth, *summer*, 1160. II. Hyacinthus *amethystinus*.
15. Hyacinth, *Spanish*, 1160. 21. H. Hyacinthus *cernuus*.
59. Star of Bethlehem, 1153. 9. H. Ornithogalum *umbellatum*.
68. Lilly, *fiery*, 1110. 3, 4, 5, 7. II. Lilium *viviparum*.
69. Moly, *yellow*, 1123. 4. H. Allium *moly*.
76. Martagon *of Pompony*, 1114. 7. H. Lilium *Pomponium*.
79. Star of Bethlehem, *spiked*, 1151. 1. H. Ornithogalum *Pyrenaicum*.

INTRODUCTION.

80. Corn flag. 1168. H. Gladiolus *communis*.
81. Martagon, *common*, 1112. H. Martagon *vulgare*, 15.
86. Martagon, *white*, 1112. Martagon *album*, 13.
100. Lilly, *white*, 1109. H. Lilium *album*.
111. Hyacinth, *dun-coloured*, 1160. 22. H. Hyacinthus *serotinus*.
113. Saffron, *meadow*, 373. Colchicum *autumnale*.

When many calendars of this kind shall be made in different places and nations in the same year, it will be easy to collect from the blowing of these sorts of flowers, and from the leafing of trees, how one climate differs from another, and why plants brought from the Southern parts seldom produce fruit with us, whereas the Northern plants succeed very well. Thus at Montpelier the spring is forwarder than at Upsal by 31 days, at London by 28, at Falconia by 6; and the winter comes on as much later in those places.

Botanists and apothecaries, whose business it is to gather plants just when they are in blow, may by this means learn at what time that may be done, and need not seek in vain at an improper season, and may farther know by their garden plants what wild ones are to be found in the fields precisely at the same time; and on the contrary.

The night frosts which so often destroy our plants, and which i imagine come to us from Lapland, may be known in the same way.

Thus the LEAD cold arising from the thaws in Lapmarck, happens at the end of the leafing season.

INTRODUCTION.

The BRASS cold from the snow melting in Lapland in the beginning of the fruiting season.

The IRON cold from the freezing on the Lapland alps in the middle of the sowing season.

These colds do not happen with us the same night as in Lapland, but arrive in about 8 days.

On these and such like calendars vulgar practical husbandry ought to be established; but the foundation hitherto not having been sufficiently well laid, this method is become so much out of use, that it is even looked upon as absurd and chimerical; nevertheless it may and ought to be carried so far, that no prudent œconomist will choose to be without such a guide, and the husbandman shall find it the surest way to regulate his affairs by, not to mention other particulars.

THE
CALENDAR
OF
FLORA,
SWEDISH and ENGLISH.

Made in the YEAR 1755.

THE MONTHS.

I. Reviving winter month from Dec. 22 to March 19.
II. Thawing month from Mar. 19 to Apr. 12.

1. SPRING.

III. Budding month from April 12 to May 9.
IV. Leafing month from May 9 to May 25.
V. Flowering month from May 25 to June 20.

2. SUMMER.

VI. Fruiting month from June 20 to July 12.
VII. Ripening month from July 12 to Aug. 4.
VIII. Reaping month from Aug. 4 to Aug. 28.

3. AUTUMN.

IX. Sowing month from Aug. 28 to Sept. 22.
X. Shedding month from Sept. 22 to Oct. 28.
XI. Freezing month from Oct. 28 to Nov. 5.

4. WINTER.

XII. Dead winter month from Nov 5 to Dec. 22.

(261)

THE
Calendar of FLORA.

I. REVIVING WINTER MONTH.
From the winter solstice to the vernal æquinox.

Dec. XII.
- xxii. *Butter shrinks and separates from the sides of the tub.*
- xxiii. Asp flower buds begin to open.

Jan. I. i. *Ice on lakes begins to crack.*
- ii. *Wooden walls snap in the night.*
  Cold frequently extreme at this time, the greatest observed was 55. 7.
- iv. * *Horse dung spirts.*
- viii. *Epiphany rains.*
- xxvi. St. *Paul's rains.*

Feb. II.
- xxii. *Very cold nights often between Feb.* 20 *and* 28, *called* Steel Nights.

---

* Note. This was explained to me by Mr. Solander, an ingenious and learned disciple of Linnæus, now in England, who says, that horse dung, in very severe frosts, throws out particles near a foot high, and that no other dung does the like.

II. THAWING MONTH.
From the first melting of the snow to the floating of ice down the rivers.

Vere novo gelidus canis cum montibus humor
Liquitur, et zephyro putris se gleba resolvit.
VIRG.

Mar. III.
- **xix.** *Eves drop towards the noontide sun.*
    Sallow, *round leaved,* flower-buds, 449.
    15. Salix *caprea,* open.
- **xx.** *Snow melts against walls.*
    LARK *begins to sing.*
- **xxii.** *Water flows by the walls.*
- **xxv.** *Roads very dirty and full of water.*

April IV.
- **i.** *Horse dung melts the ice.*
    Moss, *upright fir.* Lycopodium *selago,* 106. sheds its dust.
- **iii.** STONES *are loosened from the ice.*
- **vi.** *Hills begin to appear, the snow being melted.*
    Serpents come out of their holes.
    SPIDER, *water, frisks about.* The FLY *creeps forth.*
    GAME, *black,* 53. Tetrao tetrix.
    LAPWING, 110. Tringa vanellus, *returns.*
- **vii.** BUTTERFLY, *nettle,* Papilio urticæ, *appears in abundance.*

Some people, says Pliny, think the appearance of the butterfly the surest sign of spring, on account of the delicacy of the animal.

DUCK,

THE CALENDAR OF FLORA.

April IV.
- vii. DUCK, tame, 145. *Anas boschas sits.*
  Wild DUCK *returns.*
- x. *An inundation of snow water.*
  SWAN, 37. *Anas cygnus,* and DAKER-HEN, 58. 8. *Rallus crex, by their appearance proclaim the spring.*
  RIVERS *are unbound, and ice floats down.*

N. B. *The river at Upsal, for 70 years, has never been frozen beyond the 19th of April, according to the observation of O. Celsius, sen.*

  PIKE, 112. *Esox lucius, spawns. This fish gives over spawning when the frog begins.*
- xi. *Snow water soaks into the earth.*
  Subterraneous places are inundated.
  FROG *comes forth.*
  Winter shelters ought to be removed from garden plants, that they may not be too much drawn up.
  Hot beds for melons should be sown.

Solvitur acris hyems grata vice veris et Favonii.
HOR.

III. BUDDING MONTH.

From the return of the WHITE-WAGTAIL, *Motacilla alba, 75. 1. to the coming of the swallow; or from the first flower to the leafing of the first tree, during the whole time of the flowering of the bulbous violet.*

A Favonio veris initium notant.   CICERO.

April IV.
- xii. Hasel-nut tree, 439. *Corylus avellana.*

April IV.
- xii. COLTSFOOT, 173. Tussilago *farsara*.
- xiii. Saffron, 374. Crocus *sativus*.
  - VIOLET, *bulbous*, 1144. H. Leucoium *vernale*, 26.
  - Snow drops, 1144. H. Galanthus *nivalis*, 26.
  - WAGTAIL, *white*, 75. 1. Motacilla alba, *returns*.
  - KESTREL, 16. 16. *Falco tinunculus, returns.*
  - FROG, 247. *Rana temporaria, croaks.*
  - Saffron, 374. Crocus *sativus*, 17.
- xv. Pilewort, 246. Ranunculus *ficaria*.
  - Star of Bethlehem, *yellow*, 372. Ornithogalum *luteum*.
  - Grass, *whitlow*, 292. Draba *verna*.
  - Mezereon, 1587. H. Daphne *mezereon*.
  - TURKEY hen, 51. 3. *Meleagris gallopavo, sits.*
  - Honeysuckle, *double*, 1490. H. Lonicera *perfoliata*.
- xvi. Liverwort, *noble*, 580. H. Anemone *hepatica*, 34.
  - *The time for sowing barley at Upsal.*
- xvii. Lilly, *yellow water*, 368. 1. Nymphæa *lutea*, leaves emerge.
- xix. Asp, or *trembling* poplar, 446. 3. Populus *tremula*, 13.
  - Abele, 446. 2. Populus *alba*, 13.
  - *Hot-beds to be sown from the budding of the poplar to the leafing month.*
- xxi. SMELT, 66. 14. *Salmo eperlanus, spawns, at which time generally tempests and snowy weather at* Upsal, *and intermitting fevers very common.*

THE CALENDAR OF FLORA.

April IV.
- xxi. Hellebore, *black*, 271.1. Helleborus *viridis*.
  Willow, *round leaved*, 449. 15. Salix *caprea*.
- xxx. Crake *berries*, 444. Empetrum *nigrum*.
  Poplar, *black*, 446. Populus *nigra*.
  Bur *butter*, 179. Tussilago *petasitis*, 25.

May V.
- i. Mercury *dogs*, 138. 1. Mercurialis *perennis*.
  Polyanthus, 1083. Primula veris *hortensis*.
- iii. ANEMONE, *wood*, 259. 1. Anemone *nemorosa*, 21.
  Saxifrage, *golden*, 158. 2. Chrysosplenium *alternifolium*.
  Violet *with throat-wort leaves*, 365. 8. Viola *hirta*.
  Assarabacca, 158. 1. Asarum *Europæum*.
  Violet, *sweet*, 364. 8. Viola *odorata*, 24.
  Pepperwort, 304. Lepidium *petræum*.
  Fields are covered with verdure.

After the return of the WHEAT EAR*, 75. 1. *Motacilla* Oenanthe, *there is seldom any severe frost, and therefore the peasants in Upland have this proverb; When you see the* WHITE WAGTAIL, *you may turn your sheep into the* † *fields; and when you see the* WHEAT EAR, *you may sow your grain.*

---

* If this bird does not quit England, it certainly shifts places. For I have observed, that about harvest time they were not to be found where there were before great plenty of them.

† The sheep are housed all winter in Sweden, as Mr. Solander informs me, who gave me the translation of the Swedish proverb in the very words here printed.

Osier,

May V.

- vii. Ofier, 450. 21. Salix *viminalis*.
  Bramble, 467, 1. Rubus *fruticosus*, leafs.
  STARLING, 67. 1, *Sturnus vulgaris*, *returns*.
- viii. Moscatel, *tuberous*, 267. Adoxa *moscatellina*.
  Seeds of kitchen plants to be sown. Tender plants to be taken out of the green house.
  ELM TREE, 469. Ulmus *campestris*.
  Snow melts even in the shade.

Diffugere nives, redeunt jam gramina campis.
Arboribusque comæ   HOR.

IV. LEAFING MONTH.

The compleat leafing of trees from the bird cherry to the ash; from the coming of the swallow to the tulip.

Nunc herbæ rupta tellure cacumina tollunt;
Nunc tumido gemmas cortice palmes agit.   OVID.

May V.

- ix. SWALLOW *and* STORK *return*.
  CHERRY, BIRD, 463. Prunus *padus**. L.
  Filberd, 439. Corylus *avellana*. L.
- xi. Asp, 446. 3. Populus *tremula*, *out of blow*.
- xii. CUCKOW, 23. *Cuculus canorus*, *sings*.

---

* The letter **L.** at the end of the lines signifies that those plants came into leaf on the days marked. All the plants besides throughout this Calendar are supposed to have flowered on the days marked, unless the contrary is expressed.

THE CALENDAR OF FLORA. 267

May V.
- **xiii.** Sorrel *wood*, *281. Oxalis *acetosella*.
  BIRCH TREE, 443. Betula *alba*. L.
  Barberry bush, 465. Berberis *vulgaris*. L.
  The best time for sowing barley, and the seeds of garden plants.
  Osier, 450. 21. Salix *viminalis*. L.
- **xiv.** Spindle tree, 468. Euonymus *Europæus*. L.
  Bear's ear, 1083. H. Primula *auricula*, 12.
  Goule, or Dutch myrtle, 443. Myrica *gale*.
  Orange, *mock*, 1763. H. Philadelphus *coronarius*. L.
  Elder, *water*, 460. Viburnum *opulus*. L.
  Lilac, 1763. H. Syringa *vulgaris*. L.
  Privet, 463. Ligustrum *vulgare*. L.
  Buckthorn, *sea*, 445. Hippophae *rhamnoid*. L.
  Alder tree, 442. Betula *alnus*. L.
- **xv.** Daffodil, *wild English*, 371. 1. Narcissus *pseudonarcissus*, 19.
  Roses, *garden*.
  Elm tree, 469. Ulmus *campestris*. L.
  NIGHTINGALE, 78. *Motacilla luscinia, returns.*
  Thorn, *white*, 453. 3. Cratægus *oxyacantha*. L.
  Apple tree, 451. Pyrus *malus*. L.
  Primrose, 284. Primula *veris*, 16.
  Cherry tree, 463. Prunus *cerasus*, L.
  Thorn *buck*, 466. Rhamnus *catharticus*. L.
  Cinquefoil, *small rough*, 323. Potentilla *verna*. 16.
- **xvi.** Sallow, *round leaved*, 449. 15. Salix *caprea*. L.

Beam

THE CALENDAR OF FLORA.

May V.

- **xvi.** Beam tree, *white*, 453. Cratægus *aria* L.
  Chesnut tree, *horse*, 1683. Æsculus *hippocastanum*. L.
  Beech tree, 439. Fagus *sylvatica*. L.
  Hornbeam, 451. Carpinus *betulus*. L.
  Poplar, *black*, 446. Populus nigra. L.
- **xx.** Asp, 446. Populus *tremula*. L.
- **xxi.** Marygold, *marsh*, 272. Caltha *palustris*.
  Lime tree, 473. Tilia *Europæa*. L.
  Alder, *berry bearing*, 465. Rhamnus *frangula*. L.
  Fly, dragon, Moufet, p. 67. Libellula.
  Salmon, 63. 2. Salmo, *salar*.
  Oak tree, 440. Quercus, *robur*. L.
  Ash tree, 469. Fraxinus *excelsior*. L.
- **xxiv.** *While the ash is leafing there is scarcely any more frost: therefore green house plants ought to be brought into the open air.*

The leaden nights *happen before the leafing of the ash; from that time the summer is settled.*

There are very few flowers in this month; for nature being intent on the young offspring of the bird kind, prepares abundance of flowers, against the hatching season.

V. FLOWERING MONTH.

From the first ear of rye to its blow.
From the tulip, 1146. *Tulipa Gesneriana, to the wall pepper,* 270. 5. *Sedum acre.*

It ver, et Venus, et Veneris prænuncius ante
Pinnatus graditur zephyrus vestigia propter.
                                                    Lucret.

* Cherry,

THE CALENDAR OF FLORA.

May V.
- **xxv.** *CHERRY, BIRD, 463. Prunus *padus*.
  Currants, *black*, 456. Ribes *nigrum*.
  Jack by the hedge, 293. Erysimum *alliaria*.
  Cicely, *wild*, 207. Chærophyllum *sylvestre*.
  TULIP, 1146. H. Tulipa *Gesneriana*, 13.
  Milkwort, *287. Polygala *vulgaris*.
  Lady's Mantle, 158. Alchemilla *vulgaris*.
  ROCHE, 122. *Cyprinus rutilus, spawns*.
  First ear of rye.
- **xxvi.** Saxifrage, *white*, 364. 6. Saxifraga *granulata*.
  Ivy, *ground*, 243. Glechoma *hederacea*.
  Goldilocks, 248. Ranunculus *auricomus*.
  Pear tree, 452. Pyrus *communis*, 14.
- **xxvii.** Celandine, *greater*, 309. Chelidonium *majus*.
  Cloud berries, 260. Rubus *chamæmorum*.
  Cat's foot, 181. Gnaphalium *dioicum*.
  CRANE'S BILL, 361. 18. Geranium *sylvaticum*.
  Globe flower, 272. Trollius *Europæus*.
- **xxviii.** Cuckow flower, 299. Cardamine *pratensis*.

While the cuckow flower blows, the salmon goes up the rivers, and dragon fly comes forth.

  Thorn, *black*, 462. Prunus *spinosa* 10.
  Cherry tree, 463. Prunus *cerasus*, 10.
  Plumb tree, 462. 2, 3. Prunus *domestica*.
  Pease, *wood*, 124. 2. Orobus *tuberosus*.
  Plantain, *hoary*, 314. 3. Plantago *media*.

---

* While the bird cherry flowers, happens what is called the grey weather, between the old and new moon. I am indebted to Mr. Solander for the interpretation of this passage: it is in the original called the plenilunium cornicum, but ought to be read interlunium.

Butter-

May V.

xxx. Butterwort, * 281. Pinguicola *vulgaris*.
Lilly in the valley, 264. Convallaria *Maialis*, 21.
Bugle, *mountain*, 245. 2. Ajuga *pyramidalis*.
Rush, *hare's tail*, 436. Eriphorum *vaginatum*.
Grass, *cotton*, 435. Eriophorum *polystachyon*.
Honeysuckle, *dwarf*, 261. Cornus *Suecica*.
Whorts, *red*, 457. Vaccinium *vitis idæa*.
CROWFOOT, CRANE'S BILL, 361. 18. Geranium *sylvaticum*, 120.
Catchfly, *red German*, 340. 14. Lychnis *viscaria*.
Germander, *wild*, 282. 11. Veronica *chamædrys*.
BREAM, 116. 5. *Cyprinus brama, spawns*.

The bream sports while the bird cherry fades, and the juniper begins to blow, and when this fades the dragon fly comes out.

Nunc frondent sylvæ, nunc formosissimus annus.

June VI.

i. Fir, *spruce*, 441. Pinus *abies*, drops its male flower.
ii. PEIONY, 694. H. Pæonia fl. *simplici*, 10.
AVENS, PURPLE MOUNTAIN, 253. Geum *rivale*.
Tormentil *septfoil*, 257. 1. Tormentilla *erecta*.
APPLE TREE, 451. Pyrus *malus*.
Rye in ear.
iii. Juniper tree, 444. Juniperus *communis*.
Quicken tree, 452. Sorbus *aucuparia*.
Buck-

June VI.

iii. Buckbean, 285. Menyanthes *trifoliat.* 18.
iv. Grass, *foxtail,* 396. 1. Alopecurus *pratensis.*
v. Plantain, *ribwort,* 314. 5. Plantago *lanceolata,*

Crane's bill, *dusky,* 361. 21. Geranium *phæum.*

vii. Orchis, *male handed,* 310. 19. Orchis *latifolia.*

Gooseberry bush, 1484. H. Ribes *grossularia.*

Paris, *herb,* 264. Paris *quadrifolia.*

Trefoil, *bird's foot,* 334. 1. Lotus *corniculata.*

Bilberry bush, *great,* 457. Vaccinium *uliginosum.*

Comfrey, 230. Symphytum *officinale.*

Flax, *purging,* 362. 6. Linum *catharticum.*

Columbines, 273. Aquilegia *vulgaris.*

Violet, *dame's,* 790. H. Hesperis *matronalis.*

Flora unrobed. The pear, plumb, Scotch fir and spruce out of blow.

The summer is then in its highest beauty, when Pomona, dressed as it were in her snow-white garment, celebrates her nuptials; while the tulip, narcissus and peiony adorn the garden, the fresh shoots of the fir illuminate the woods, and the juniper sheds its impregnating vapour.

Christopher, *herb,* 661. Actæa *spicata.*

Grass, *melic,* 403. 6. Melica *nutans.*

Bramble, *stone,* 261. Rubus *saxatilis.*

Crowfoot, *bulbous,* 247. 2. Ranunculus *bulbosus.*

Hawk-

June VI.

vii. Hawkweed, *dandelion*, 245. Leontodon *hispidum*.

viii. Barberry, 465. Berberis *vulgaris*.
  LILAC, 176. H. Syringa *vulgaris*.
  The meadows glow with crowfoots.

ix. Crane's bill, *dovefoot*, 359. 11. Geranium *molle*.

x. Flower de luce, *yellow water*, 374. Iris *pseud-acorus*.

xi. Crane's bill, *crowfoot*, 360 17. Geranium *pratense*.
  Bell flower, *lesser round-leaved*, 277. 5. Campanula *rotundifolia*.
  Camomile, *Roman*, 189. Matricaria *chamomilla*.
  Cinquefoil, *shrubby*, 256. 4. Potentilla *fruticosa*.

xii. Beam tree, *white*, 453. Cratægus *aria*.
  Vetch, *kidney*, 325. 1. Anthyllis *vulneraria*.
  Henbane, 274. Hyoscyamus *niger*.

xiii. Catchfly, *white*, 340. 11. Silene *nutans*.
  Avens, 253. 1, 2. Geum *urbanum*.
  Adonis, *flower*, 251. Adonis *annua rubra*.

The single peiony fades before the double blows, as is the case of rhubarb and rhapontic.

xiv. Lilly, *yellow water*, 368. 1. Nymphæa *lutea*.
  Orchis, *female handed*, 381. 20. Orchis *maculata*.
  Robert, *herb*, 358. 6. Geranium *robertianum*.
  Cinquefoil, *upright bastard*, 255. Potentilla *rupestris*.
  Campion, *white*, 339. 8. Lychnis *dioica*.
  Elder,

THE CALENDAR OF FLORA.

June VI.
- xiv. Elder, *water*, 460. Viburnum *opulus*.
- xv. Rose, *pimpernel*, 454. 3. Rosa *spinosissima*.
  Briar, *sweet*, 454. 2. Rosa *eglanteria*.
  Thorn, *buck*, 466. Rhamnus *catharticus*.
  Orchis, *lesser butterfly*, 380. 18. Orchis *bifolia*.
- xvi. Grass of Parnassus, 355. 1. Parnassia *palustris*.
  Grass, *marsh goose*, 255. 3. Galium *uliginosum*.
  Lilly, *white water*, 368. 3. Nymphæa *alba*.
  Tansey, *wild*, 256, 5. Potentilla *anserina*.
  Alder, *berry bearing*, 465. 1. Rhamnus *frangula*.
  PEIONY, DOUBLE, 693. H. Pæonia *plena*, 13.
  Rampions, 274. 4. Campanula *patula*.
  Rhapontic, 170. H. Rheum *rhaphonticum*.
- xvii. Pink, *meadow*, 338. Lychnis *flos cuculi*.
  Valerian, *great wild*, 200. 1. Valeriana *officinalis*.
  Vetch, *chichling*, 320. 5. Lathyrus *palustris*.
  Daisy, *great, or ox eye*, 184. Chrysanthemum *leucanthem*.
- xvii. Eyebright, *284. Euphrasia *officinalis*.
  Spear wort, *lesser*, 250. 7. Ranunculus *flammula*.
  Groundsel, 178. 1. Senecio *vulgaris*.
  Thorn, *white*, 453. Crætagus *oxyacantha*.
- xviii. Lilly, *bulbous*, 1110. H. Lilium *bulbiferum* to VII. 4.
  Saxifrage, *burnet*, 213. 1, 2. Pimpinella *saxifraga*.

T Tor-

June VI.

xviii. Tormentil *cinquefoil*, 255.2.   Potentilla *argentea*.

Grals, *quaking*, 412. Briza *media*.

Nightshade, *deadly*, 265.   Atropa *belladonna*.

Rye, *winter*, 388.1. Secale *hybernum*.

Winter rye flowers generally from the first blow of wall pepper, 270.5. Sedum acre, to the first blow of the rose bay willow herb, 310.1. Epilobium angustifolium, with the bulbous lilly, beginning oftentimes four days before the solstice.

VI. FRUITING MONTH.

During the whole blow of the wall pepper.   From the yellow to the red day lilly.

———Revoluta ruebat,
Matura jam luce dies.   Virg.

xx. Pepper, wall, 207.5. Sedum *acre*.

Orchis, *frog*, 381.22. Satyrium *viride*.

Cistus, *dwarf*, 341. Cistus *helianthem*.

Lilly, yellow day, 1191. H. Hemerocallis *flava*.

Cinquefoil, *marsh*, 256.2. Comarum *palustre*.

xxi. *After the solstice trees scarcely grow, and therefore hedges should then be clipped. The highest degree of heat with us is hardly above 54.8. within the tropics the heat is not much above 57.6. nor below 40.8. the heat of a hatching hen between 63 and 68.4. a heat above 72 destroys the embryo.*

Brass

THE CALENDAR OF FLORA.

June VI.
- xxi. BRASS NIGHTS *from the thaw of the highest mountains.*
- xxii. Orchis, *fly*, 379.13. Ophrys *insectifera myodes.*

  Blue bottle, 198. Centaurea *cyanus.*

  Vetch, *great tufted wood*, 322.4. Vicia *sylvatica.*

  Dropwort, 259. Spiræa *filipendula.*

  Thistle, *gentle*, 193.2. Carduus *heleniodes.*

  Loose strife, *yellow*, 283.3. Lysimachia *thyrsiflora.*

  Self heal, 283. Prunella *vulgaris.*

  Gentian, *vernal dwarf*, 275.4. Gentiana *campestris.*

- xxiii. Mayweed, *stinking*, 185.4. Anthemis *cotula.*

  Yarrow, 183.1. Achillea *millefolium.*

  WILLOW HERB, *rose bay*, 310. Epilobium *angustifolium.*

  Moonwort, 128.1. Osmunda *lunaria.*

  Liquorice, *wild*, 326.1. Astragalus *alpinus.*

  Knapweed, *great*, 198.1. Centaurea *scabiosa.*

  Vetch, *tufted*, 322.3. Vicia *cracca.*

  Nightshade, *woody*, 265.1,2. Solanum *dulcamara.*

  Golden rod, 176.1. Solidago *virgaurea.*

  ORANGE, MOCK, 1763. H. Philadelphus *coronarius*, 14.

  Sweet-william, 991.2. H. Dianthus *barbatus.*

  POPPY, *Turkey*, Papaver *orientale*, 13.

  Flax, *perennial blue*, 362.3. Linum *perenne.*

Dewberry

June VI.

xxiii. Dewberry bush, 467.3. Rubus *cæsius*.
xxv. Nettle, *hedge*, 237. Stachys *sylvatica*.
  Spiked willow of Theophrastus, 1699.
   H. Spiræa *salicifolia*.
  Lilac out of blow.
xxvi. Willow herb, *hooded*, 244.1. Scutellaria *galericulata*.
  Willow herb, *great smooth leaved*, 311.4. Epilobium *montanum*.
  Twayblade, 385.1. Ophrys *ovata*.
  Strawberries ripening.
xxvii. Hawkweed, *Hungarian*, 167.17. Hypochæris *maculata*.
  Medic, *yellow*, 333.1. Medicago *falcata*.
  Parsley, *great bastard*, 219.2. Tordylium *latifolium*.
xxviii. Toad flax, *yellow*, *281.1. Antirrhinum *linaria*.
  Grass, *sea dog*, 390.1. Elymus *arenaria*.
  Bryony, *white*, 261.1,2. Bryonia *alba*.
  Campion, *wild purple*, 341.17. Silene *armeria*.
xxix. Marygold, *corn*, 182.1. Chrysanthemum *segetum*.
  Heath, *Dutch or besom*, 471.4. Erica *tetralix*.
  Bilberry bush, 457.2. Vaccinium *myrtill*.
  Berries ripe.
  Pease, *everlasting*, 319.1. Lathyrus *latifolius*.
  Throatwort, *little*, 277.3. Campanula *glomerata*.
  Feverfew, 187.1. Matricaria *parthenium*.
  PEACOCK *moults*.

THE CALENDAR OF FLORA.

June VI.
- xxix. Ox-eye, 183.1. Anthemis *tinctoria*.
  Sneezewort, 183. Achillea *ptarmica*.
  Rupturewort, 160.1. Herniaria *glabra*.
  Hawkweed, *succory leaved*, 166.12. Crepis *biennis*.
  PINK, MAIDEN, 335.1. Dianthus *deltoides*.
  SCABIOUS, FIELD, 191.1. Scabiosa *arvensis*.
  St. John's wort, *large flowered*, 1017.1. H. Hypericum *ascyron*.
  Elder tree, 461.1. Sambucus *nigra*.
  Woad, 367.1. Isatis *tinctoria*, out of blow.

July VII.
- i. Willow herb, *purple spiked*, 367.1. Lythrum *salicaria*.
  Parsnep, *cow*, 205.1,2. Heracleum *sphondylium*.
  Bindweed, *small*, 275.1. Convolvulus *arvensis*.
  Knapweed, 198.2. Centaurea *nigra*.
  Mullein, *white flowered*, 287.2. Verbascum *lychnit. alba*.
  Rampions, 277.4. Campanula *ranunculus*.
- ii. Throatwort, *giant*, 276. Campanula *latifolia*.
  Asphodel, *Lancashire*, 375.1. Anthericum *ossifragum*.
  Mullein, *black*, 288.4. Verbascum *nigrum*.
  Rue, *meadow*, 203.1. Thalictrum *flavum*.
  Hellebore, *bastard*, 383.1. Serapias *hellebor. palustr*.

The hottest days.

Bindweed, *great*, 275. Convolvulus *sepium*.
Willow herb, *yellow*, 282.1. Lysimachia *vulgaris*.

Moneywort.

THE CALENDAR OF FLORA.

July VII.

- iv. Moneywort, 283.1. Lysimachia *nummularia*.

    Foxglove, purple, 283.1. Digitalis *rubra*.

    Meadow sweet, 259.1. Spiræa *ulmaria*.

    Cockle, 338.5. Agrostema *githago*.

    Speedwell, *spiked male*, 279.2. Veronica *spicata*.

    Grass, *soft tufted meadow*, 404.14. Holcus *lanatus*.

    Primrose, *evening or tree*, 862. Oenothera *biennis*.

- vi. Yarrow, 183.1. Achillea *millefolium*.

    Bedstraw, yellow lady's, 224.1. Galium *verum*.

    Agrimony, 202.1. Agrimonia *eupatoria*.

    Throatwort, *great*, 276. Campanula *trachelium*, 25.

    St. John's wort, 342.1. Hypericum *perforatum*.

    St. John's wort, *tutsan*, 343.4,5. Hypericum *hirsutum*.

    Spearwort, *great*, 250.8. Ranunculus *lingua*.

    Carrot, 218. Daucus *carota*.

    Stone crop, *yellow*, 269.1. Sedum *rupestre*.

    Gladdon, *stinking*, 375.3. Iris *fœtidissima*.

    Knapweed, 198.2. Centaurea *jacea*.

    Hops, 137.1. Humulus *lupulus*.

    Rest harrow, 332.1. Ononis *spinosa*.

    Parsley, *Scotch sea*, 214. Ligusticum *Scoticum*.

- vii. Briar, or dog rose, 454.1. Rosa *canina*.

    Rose, *white*, 1373.23. H. Rosa *alba*.

    Rose, *French*, Rosa *Gallica*.

THE CALENDAR OF FLORA.

July VII.

vii. *The late roses now begin to blow.*
Hay harvest begins with the lime tree, clover being out of blow, and yellow rattle or coxcomb, *284. *shedding its seeds.*
Burnet, 203.2. Sanguiforba *officinalis.*
Poppy, *wild,* 308.1. Papaver *somniferum.*
Sneezewort, 183.1. Achillea *ptarmica.*
Lilly, *yellow day, out of blow.*
Chervil, *wild,* 207.1. Chærophyllum *sylvestre, out of blow.*
Barley every where in ear.
Pease ripe.
Cherries ripe.
Beginning of hay harvest.
Motherwort, 239.1. Leonurus *cardiaca.*

viii. Pink, *Deptford,* 337.1. Dianthus *armerius.*
Orange, *mock, out of blow.*
Bilberries ripe.

ix. Thistle, *musk,* 193.1. Carduus *nutans.*
Burdock, 196. Arctium *lappa.*
Horehound, *base,* 339.1. Stachys *Germanica.*
Hemp, 138. Cannabis *sativa.*
351.13. Frankenia *pulverulenta.*
Red currants ripe.

x. Mugwort, 190.1. Artemisia *vulgaris.*
Thistle, *ivy leaved sow,* 162.5. Prenanthes *muralis.*
Marjoram, *wild,* 236.1. Origanum *vulgare.*
Horehound, *stinking,* 244.1. Ballota *nigra.*
Basil, *great wild,* 239. Clinopodium *vulgare.*
Pine, *ground,* 244. Teucrium *chamæpitys.*

Betony,

July VII.
- x. Betony, *water*, *283.1. Scrophularia *aquatica*.
  Nightshade, *enchanters*, 289. Circæa *canadensis*.
  Clover, 328.4. Trifolium *pratense*, *out of blow*.
- xi. Thistle, *tree sow*, 163.7. Sonchus *arvensis*.
  LIME TREE, 473.1,2,3. Tilia *Europea*, *out of blow*.
  Marjoram, *wild*, 236. Origanum *vulgare*.

Festinat decurrere velox flosculus æstatis.

VII. RIPENING MONTH.

From the white stonecrop, 271.7. Sedum album, or the red day lilly, to the devil's bit, 191.3. Scabiosa succisa.

- xii. SUCCORY, WILD, 172. Cichorium *intybus*.
  Willow herb, *great hairy*, 311.2. Epilobium *hirsutum*.
  Langue de boeuf, 166.13. Picris *echioides*.
  Woodbind, 456. Lonicera *periclymenum*.
  Mallow, *vervain*, 252. Malva *alcæa*.
  St. John's wort, *large flowered*, 1017. H. Hypericum *ascyron*, *out of blow*.
- xiii. Fleabane, *small*, 174.2. Inula pulicaria.
  Pepperwort, 304.1. Lepidium *latifolium*.
- xiv. STONECROP, WHITE, 271.7. Sedum *album*, *out of blow*.
- xv. Agrimony, *hemp*, 179. Eupatorium *cannabinum*.

Tansey,

THE CALENDAR OF FLORA.

July VII.
- **xv.** Tanfey, 188. Tanacetum *vulgare*.
  Golden rod, *white*, 175. 1. Erigeron *canadenfe*.
  Saw-wort, 196.1. Secratula *tinct.præalt*.
  Mint, *red*, 232.5. Mentha *gentilis*.
  Mint, *long leaved horfe*, 234.5. Mentha *fpicata*.
  Pepper, *wall*, 270. 5. Sedum *acre, out of blow*.
  CUCKOW *is filent*.
- **xvi.** Hawkweed, *bufhy*, 168.3. Hieracium *umbell*.
  LILLY, RED DAY, 1191.2. H. Hemerocallis *fulva*.
  All the marvels of Peru, 398. H. Mirabiles.
  The height of hay harveft.
- **xvii.** Thiftle,*fpear*,194.8.Carduus*lanceolatus*.
  ELDER, DWARF, 461.4. Sambucus *ebulus*.
- **xviii.** Touch me not, 316.1. Impatiens *noli tangere*.
  Saffron *meadow*, 373.1. Colchicum *autumnale. Leaves fall.*
  Teafel, *wild and manured*, 192.3. Dipfacus *fullonum*.
- **x.** LILLY, WHITE, 1109. H. Lilium *candidum*.
  Teafel,*fmall wild*,192.3.Dipfacus*pilofus*.
- **xxi.** Thiftle, *marfh tree fow*, 163.8. Sonchus *paluftris*.
- **xxii.** Soapwort, 339.6,7. Saponaria *officinalis*.
  Grafs, *Effex cocks foot*, 393.4. Dactylis *cynofuroides*.
  Knapweed, great, 198.1. Centaurea *fcabiofa*.
  Spikenard,

July VII.

- **xxiii.** Spikenard, *plowman's*, 179.1. Conyza *squarrosa*.
- **xxiv.** Elecampane, 176.1. Inula *helenium*.
- **xxv.** Fleabane, *middle*, 174.1. Inula *dysenter*.
- **xxvi.** Violet, *Calathian*, 274.1. Gentiana *pneumonanthe*.
  Baum, 570. H. Melissa *officinalis*.
- **xxvii.** Thistle, *great, soft or gentle*, 193.3. Serratula *alpin. lat.*

Aug. VIII.

- **i.** Chickweed, *berry bearing*, 267.1. Cucubalus *baccifer*.
  Orpine, 269.1. Sedum *telephium*.

Mirabar celerem fugitiva æstate rapinam,
Et dum nascuntur consenuisse rosas.

VIII. REAPING MONTH.

From the devil's bit to the blow of the meadow saffron.

- **iv.** DEVIL'S BIT, 191.3. Scabiosa *succisa*.
  Rye harvest.
  Winter rye has for many years ripened with the first blow of the devil's bit, in the garden at Upsal.
  BIRDS OF PASSAGE, *after having celebrated their nuptials in the vernal months, and feasted on the summer fruits, now prepare for departing.*
- **vii.** Rose, *French*, Rosa Gallica, *out of blow.*
- **xiv.** Wormwood, 188.1. Artemisia *campestris*.
- **xvi.** *Barley harvest.*
- **xxvi.** Lilly, *red, out of blow.*

THE CALENDAR OF FLORA.

*IX. SOWING MONTH.

From the first blow of the meadow saffron to the departure of the swallow.

Pomifer autumnus fruges effudit, et mox
Bruma recurrit iners.

Aug. VIII.
  xxviii. SAFFRON, MEADOW, 373.1. Colchicum *autumn.*

This plant ought to admonish gardeners to put Indian plants under shelter, as the iron nights are near. The IRON NIGHTS, *as they are called with us, generally happen between August* 17 *and* 29, *and destroy tender plants.*

After the blow of the meadow saffron, we have storms that shake off ripe seeds.

      Fern, female, 124.1. Pteris *aquilina*, grows yellow in the woods after the first cold nights.

  xxxi. *A gentle frost that scarcely did any damage.*
Sept. IX.
      i. Adonis flower, 251.1. Adonis *annua ser.*
      Mulberry tree, 1429. H. Morus *nigra*, grows pale.
  iv. Fig tree, 1431. H. Ficus *carica* grows pale.
  vi. Wormwood, *sea*, 188.2. Artemisia *maritima*.
      Travellers joy, 258.1. Clematis *vitalba*.
  xi. *The frost has destroyed southern plants.*
  xii. *The frost milder.*
  xiv. *Seeds to be gathered.*

---

* By sowing, in this place, is meant not man's but nature's.

Sept. IX.
> xvii. Sampire, *golden flowered*, 174.1. Inula
> *crithmoid*.
>> * SWALLOW *goes under water.*
>> WAGTAIL, *white, departs.*

Nos quoque floruimus, fed flos fuit ille caducus.

X. SHEDDING MONTH.
From the first fall of the leaves of trees to the last.

> xxii. Orach, *sea*, 152.8. Artiplex *maritima*.
>
> *Leaves of trees are changed, of the oak, maple, robinia caragana, elm, lime, to a yellow; of the spin tree to a brown; of the quicken tree and sumach to a red colour.*
>
>> *Leaves of the oak dry and yellow.*
>
> xxiv. ‖ *Leaves of the maple begin to fall in the night.*
> xxv. *Hoar frost.*
> xxvi. *Leaves of the robinia caragana fall.*
>> Sycamore *stripped of its leaves.*

Oct. X.
> iv. Cherry, *bird, stripped of its leaves.*
> v. *A storm.*

---

* Adamson in the account of his voyage to Senegal, p. 121. says, that Oct. 1749, European swallows lodged on the vessel in which he went from Goree to Senegal, and that they are never seen there but at this time of the year, along with quails, wagtails, kites, and some other birds of passage, and do not build nests there. This testimony seems to take away all doubts about this long contested point.

‖ Geminus, either from himself or Democritus, which is much the same, as Rome and Abdera were nearly in the same latitude, says that the leaves of trees began to drop the fourth of Scorpio, which answers to October 28.

Green

THE CALENDAR OF FLORA.

Oct. X.

- vi. *Green leaves of the ash fall. When they fall, southern plants ought to be put under shelter.*
- vii. Elm *is stripped.*
  A storm.
  Frost.
- xii. *The leaves of the lime tree fall. Green houses ought to be shut.*
- xiii. *The asp tree still in leaf.*
- xiv. *Ice.*
- xvii. *Hasel nut tree stripped.*
- xxv. Abele, 446.2. Populus *alba, stripped.*
- xxvi. Saffron, *meadow, just out of blow.*
  Poplar, *black, stripped.*
- xxvii. *Summer ended.*
- xxviii. *Sallows only in leaf.*

Vernantesque comas tristis ademit hyems.
          PETRON.

XI. FREEZING MONTH.

From the last shedding of leaves to the last green plant.

Iva *annua.*

Nov. XI.

- ii. *Alternate snow and frost.*
- v. Milleria *quinqueflora.*
  Thaw with rain.
  The earth covered with snow.
  Rivers are frozen.
  Snow with a thaw.
  Firm snow.
  Thaws again.

Ditches

Ditches filled with water.
Winter thoroughly settled.

Heu quam cuncta abeunt celeri mortalia cursu.

XII. DECLINING WINTER MONTH.
From the last green plant to the winter solstice.

Nov. XI.
    v. *Mosses and lichens only flourish.*
      Thermometer, gr. 34. 2.
   vi. *Thaws.*
  xx. *Cold changeable weather.*

Quælibet orta cadunt, et finem cæpta videbunt.

THE
CALENDAR of FLORA.

BY

BENJAMIN STILLINGFLEET.

Made at STRATTON in NORFOLK,
Anno 1755. Latitude 52° 45.

MARKS EXPLAINED.

b signifies buds swelled.
B - - - - buds beginning to open.
f - - - - - flowers beginning to open.
F - - - - flowers full blown.
l - - - - - leaves beginning to open.
L - - - - leaves quite out.
r. p. - - - fruit nearly ripe.
R. P. - - fruit quite ripe.
E - - - - emerging out of the ground.
D - - - flowers decayed.

THE CALENDAR of FLORA.

I. MONTH.

Reviving nature seems again to breath,
As loosen'd from the cold embrace of death.

Jan. 5. Rosemary, 515. H. Rosmarinus *officinal.* f.
11. Honeysuckle, 458. Lonicera *periclymenum*, l.
23. Archangel, *red*, 240.2. Lamium *purpureum*, F.
Hasel nut tree, 439. Corylus *avellana*, f.
Honeysuckle, 458. Lonicera *periclymenum*, L.
Laurustinus, 1690. H. Viburnum *tinus*, F.
Holly, 466. Ilex. *aquifolium*, f.
26. Snow drops, 1144. H. Galanthus *nivalis*, F.
Chickweed, 347.6. Alsine *media*, F.
Spurry, 351.7. Spergula *arvensis*, F.
Daisy, 184. Bellis *perennis*, F.

II. MONTH.

Love's pleasing ferment gently now begins
To warm the flowing blood.

Feb. 4. WOOD LARK, 69.2. *Alauda arborea*, sings.
Elder tree, 461. Sambucus *nigra*, f.
12. ROOKS, 39.3. *Corvus frugilegus*, begin to pair.
GEESE, 136.1. *Anas, anser*, begin to lay.
*WAGTAIL WHITE, 75.1. *Motacilla alba*, appears.
16. THRUSH, 64.2. *Turdus musicus*, sings.
†CHAFFINCH, 88. *Fringilla cælebs*, sings.
20. *Thermemeter,* 11. *Highest this month.*
Thermometer, -2. *Lowest this month.*
22. PARTRIDGES, 57. *Tetrao perdix*, begin to pair.
Hasel tree, 439. Corylus *avellana*, F.
25. Gooseberry bush, 1484. H. ⎫
Ribes *grossularia,* l.   ⎬ both young
Currant, *red,* 456.1. Ribes ⎪ plants.
rubrum, l.   ⎭

Thermometer from the 19th to the 25th, between 0 and -1 with snow.

Wind during the latter half of the month between E. and N.

---

* The wagtail is said by Willughby to remain with us all the year in the severest weather. It seems to me to shift its quarters at least, if it does not go out of England. However, it is certainly a bird of passage in some countries, if we can believe Aldrovandus, the author of the Swedish Calendar, and the author of the treatise De Migrationibus Avium. Linnæus observes, S. N. Art. Motacilla, that most birds which live upon insects, and not grains, migrate.

† Linnæus says, that the female chaffinch goes to Italy alone, thro' Holland; and that the male in the spring, changing its note, foretells the summer: and Gesner, ornithol. p. 388. says that the female chaffinch disappears in Switzerland in the winter, but not the male.

III.

III. MONTH.

Winter still ling'ring on the verge of spring,
Retires reluctant, and from time to time
Looks back, while at his keen and chilling breath
Fair Flora sickens.

March
2. ROOKS, 39.3. *Corvus frugilegus, begin to build.*
   Thermometer, 10.
4. THRUSH, 64.2. *Turdus musicus, sings.*
   Thermometer, 11.
5. DOVE, RING, 62.9. *Columba palumbus, cooes.*
7. *Thermometer,* 0. *Lowest this month.*
11. Sallow, Salix, F.
    Laurustinus, 1690. H. Viburnum *tinus*, l.
    *BEES, *Apis mellifera, out of the hive.*
    Laurel, 1549. H. Prunus *laurocerasus*, L.
    Bay, 1688. H. Laurus *nobilis*, l.
20. *Vernal equinox.*
21. Grass, *scurvy,* 302.1. Cochlearia *officinalis,* F.
    Asp, 446.3. Populus *tremula,* F.
26. Speedwell, *germander,* 279.4. Veronica *agrestis,* F.
    Alder, 442. Alnus *betula,* F.
28. Violet, *sweet,* 364.2. Viola *odorata,* F.
    Parsnep, *cow,* 205. Heracleum *sphondylium,* E.
    Pilewort, 296. Ranunculus *ficaria,* F.

* Pliny, nat. hist. lib. 11. §. 5. says, that bees do not come out of their hives before May 11. and seems to blame Aristotle for saying that they come out in the beginning of spring, i. e. March 12,

March
28. *Thermometer*, 25.50. *Highest this month.*
29. Cherry tree, 463. Prunus *cerasus*, B.
Currant bush, 456.1. Ribes *rubrum*, B.
Primrose, 284.1. Primula *veris*, F.
Yew tree, 445. Taxus *baccata*, F.
Elder, *water*, 460. Viburnum *opulus*, B.
Thorn, haw, 453.3. Cratægus *oxyacantha*, B.
Larch tree, 1405. H. Pinus *larix*, B.
Hornbeam, 451. Carpinus *ostrya*, B.
Tansy, 188. Tanacetum *vulgare*, E.

IV. MONTH.

————Airs, vernal airs,
Breathing the smell of grove and field, attune
The trembling leaves.   MILTON.

April 1. Chesnut, *horse*, 1683. Æsculus *hippocastanum*, B.
Birch, 443. Betula *alba*, L.
Willow, *weeping*,   Salix *Babylonica*, L.
Elm tree, 468. Ulmus *campestris*, F.
Quicken tree, 452.2. Sorbus *aucuparia*, f.
Apricot, 1533. H. Prunus *Armeniaca*, F.
Narcissus, *pale*, 371.2. Narcissus *pseudonar.*
8. Holly, 466.1. Ilex *aquifolium*, f.
Bramble, 467.1. Rubus *fruticosus*, L.
Rasberry bush, 467.4. Rubus *idæus*, L.
Currants, *red*, 456. Ribes *rubrum*, F.
Dandelion, 170.1. Leontodon *taraxicum*, E.

Cleavers,

THE CALENDAR OF FLORA.

April 3. Cleavers, 225. Galium *aparine*, E.
  4. Laurustinus, 1690. H. Viburnum *tinus*, F.
   APPLE TREE, 451.1,2. Pyrus *malus*, B.
   Orpine 269.1. Sedum *telephium*, B.
   Briar, 454.1. Rosa *canina*, L.
  6. Gooseberry, 1489. H. Ribes *grossularia*, f.
   Maple, 470.2. Acer *campestre*, B.
   Peach, 1515. H. Amygdalus *Persica*, L. et F.
   Apricot, 1533. H. Malus *Armeniaca*, L.
   Plumb tree, 462. Prunus *præcox*, L.
   Pear tree, 452. Pyrus *communis*, B.
   * SWALLOW, 71.2. Hirundo *urbica*, returns.
  7. Filberd, 439. Corylus *avellana*, L.
   Sallow,   Salix, L.
   Alder, 442. 1. Betula *alnus*, l.
   Lilac, 1763. Syringa *vulgaris*, l.
   Oak, 440.1. Quercus, *robur*, f.
   Willow, *weeping*, Salix *Babylonica*, b.
  8. Juniper, 444. Juniperus *communis*, b.
  9. Lilac, 1763. Syringa *vulgaris*, b.
   Sycamore, 470. Acer *pseudoplatanus*, L.
   Wormwood, 188.1. Artemisia *absinthium*, E.
   † NIGHTINGALE, 78. Motacilla *luscinia*, *sings*.

          Auricula,

---

* According to Ptolemy, Swallows return to Ægypt about the latter end of January.

† From morn 'till eve, 'tis music all around;
Nor dost thou, Philomel, disdain to join,
Even in the mid-day glare, and aid the quire.
But thy sweet song calls for an hour apart,
When solemn Night beneath his canopy,
Enrich'd with stars, by Silence and by Sleep

April

9. Auricula, 1082. H. Primula *auricula*, b.
10. Bay, 1688. H. Laurus *nobilis*, L.
    Hornbeam, 451. Carpinus *betulus*, b.
    Willow, *white*, 447.1. Salix *alba*, b.
    BEES *about the male fallows.*
    Feverfew, 187.1. Matricaria *Parthenium*, E.
    Dandelion, 170.1. Leontodon *taraxicum*, E.
    Hound's tongue, 226.1. Cynoglossum *officinale*, E.
    Elm, 468. Ulmus *campestris*, l.
    ANEMONE, *wood*, 259. Anemone *nemorosa*, F.
    Jack in the hedge, 291. Erysimum *alliaria*, E.
    Quince tree, 1452. H. Pyrus *cydonia*, L.
11. Elder, *water*, 460. Viburnum *opulus*, L.
    Alder, *berry bearing*, 465. Rhamnus *frangula*, L.
12. Acacia, 1719. H. Robinia *acacia*, l.
    Mulberry tree, 1429. H. Morus *nigra*, l.
    Lime tree, 473.1,2,3. Tilia *Europæa*, l.
    Mercury, *dogs*, 138.1. Mercurialis *perennis*, F.
    * Elm, *wych*, 469.4. L.
    Ragweed, 177. Senecio *jacobæa*, E.

Attended, fits and nods, in awful state;
Or when the Moon in her refulgent car,
Triumphant rides amidst the silver clouds,
Tinging them as she passes, and with rays
Of mildest lustre gilds the scene below;
While zephyrs bland breath thro' the thickening shade,
With breath so gentle, and so soft, that e'en
The poplar's trembling leaf forgets to move,
And mimic with its sound the vernal shower;
Then let me sit, and listen to thy strains, &c.

Linnæus does not seem to know this species of elm.

THE CALENDAR OF FLORA.

April
13. Laburnum, 1721. Cytisus *laburnum*, f.
Strawberry, 254. Fragaria *vesca*, F.
Quicken tree, 452.2. Sorbus *aucuparia*, L.
Sycomore, 470. Acer *pseudoplat.* L.
Laurel, 1549. H. Prunus *laurocerasus*, L.
Gooseberry bush, 1484. H. Ribes *grossularia*, F.
Currant bush, 456.1. Ribes *rubrum*, F.
Mallow, 251.1. Malva *sylvestris*, E.
Hornbeam, 451. Carpinus *betulus*, L.
14. Flixweed, 298.3. Sisymbrium *sophia*, E.
Apple tree, 451. Pyrus *malus*, L.
Hops, 137.1. Humulus *lupinus*, E.
Plane tree, 1706. H. Platanus *orientalis*, b.
Walnut tree, 438. Juglans *regia*, f.
BITTERN, 100.11. Ardea *stellaris makes a noise*.
15. Vine, 1613. Vitis *vinifera*, B.
Turneps, 204.1. Brassica *rapa*, F.
16. Abele, 446.2. Populus *alba*, B.
Chesnut, 138.2. H. Fagus *castanea*, B.
Ivy, ground, 243. Glechoma *hederacea*, F.
Fig tree, 1431. Ficus *carica*, b.
Apricots and peaches out of blow.
RED START, 78.5. Motacilla *Phænicurus*, *returns*.
Tulip tree, 1690. H. Liriodendron *tulipifera*, B.
Plumb tree, 462. Prunus *domestica*, F.
Sorrel, *wood*, *281.1,2. Oxalis *acetosella*, F.
Marygold, *marsh*, 272. Caltha *palustris*, F.
Laurel, *spurge*, 465. Daphne *laureola*, F.
17. Jack in the hedge, 291.2. Erysimum *alliaria*, F.

296 THE CALENDAR OF FLORA.

April
17. Willow, *white*, 447. 1. Salix *alba*, L. et F.
Cedar, 1404. H. Pinus *cedrus*, l.
Elder, *water*, 460.1. Viburnum *opulus*, f.
Abele, 446.2. Populus *alba*, L.
*CUCKOW, 23. Cuculus *canorus*, *sings*.
18. Oak, 440.1. Quercus, *robur*, l. F.
Thorn, *black*, 462.1. Prunus *spinosus*, B.
Pear tree, 452. Pyrus *communis*, f.
Mulberry tree, 1429. H. Morus *nigra*, B.
Violet, *dog*, 364.3. Viola *canina*, F.
Lime tree, 413.1,2,3. Tilia *Europæa*, L.
Nightshade, 265. Atropa *belladonna*, E.
Cherry tree, 463.1. Prunus *cerasus*, F.
Ash tree, 469. Fraxinus *excelsior*, f.
Maple, 470. Acer *campestre*, L.
Broom, 474. Spartium *scoparium*, b.
Chesnut, 138.2. Fagus *castanea*, L.
Fir, *Scotch*, 442. Pinus *sylvestris*, b.
Cuckow flower, 299. Cardamine *pratensis*.
20. *Thermometer* 42. *the highest this month.*
21. Walnut tree, 438. Juglans *regia*, L.
Plane tree, 1706. H. Platanus *orientalis*, L.
Fir, *Weymouth*, 8. dend. Pinus *tæda*, B.
Acacia, 1719. H. Robinia *pseudo-acacia*, L.
Fig tree, 1431. H. Ficus *carica*, L.
Wall flower, 291. Cheiranthus *cheiri*, F.
Poplar, *black*, 446.1. Populus *nigra*, L.
Beech tree, 439.1. Fagus *sylvatica*, L.
22. Fir, *balm of Gilead*,   Pinus *balsamea*, l. et f.

---

* Aristophanes says, that when the cuckow sung the Phœnicians reaped wheat and barley. Vid. Aves.

Young

THE CALENDAR OF FLORA.

April
22. *Young Apricots.*
    Fir, *Scotch*, 442. Pinus *sylvestris*, f.
    Ash, 469. Fraxinus *excelsior*, F. et L.
    Broom, 474. Spartium *scoparium*, L.
    Poplar, *Carolina*. L.
    Meadow sweet, 259. Spiræa *ulmaria*, E.
    Fig tree, 1431. H. Ficus *carica, fruit formed.*
    Tormentil, 257.1. Tormentilla *erecta*, E.
    Phyllerea, 1585. H. Phyllerea *latifolia*, F.
    Thorn, *evergreen*, 1459. H. Mespilus *pyracantha*, F.
    Rosemary, 515. H. Rosmarinus *officinalis*, F.
    Campion, *white*, 339.8. Lychnis *dioica*, F.
    Buckbean, 285.1. Menyanthes *trifol.* F.
    Furze, *needle*, 476.1. Genista *Anglica*, F.
    Stitchwort, 346.1. Stellaria *holostea*, F.
23. Crab tree, 451.2. Pyrus *malus sylv.* F.
    Apple tree, 451.1. Pyrus *malus*, f.
    Robert, *herb*, 358. Geranium *Robertian*, F.
    Fieldfares, 64.3, Turdus *pilaris, still here.*
24. Broom, 474. Spartium *scoparium*, F.
    Mercury, 156.15. Chenopodium *bonus henr.* F.
    Yew tree, 445. Taxus *baccifera*, L.
    Holly, 466.1. Ilex *aquifolium*, B.
    Furze, 475. Ulex *Europæus*, l.
    Agrimony, 202. Agrimonia *eupator*, E.
25. Sycomore, 470. Acer *pseudoplat.* F.
    Hornbeam, 451. Carpinus *betulus*, F.
    Asp, 446. Populus *tremula*, l.
    Spurge, *sun*, 313.8. Euphorbia *peplus*, F.
    Elder tree, 461.1. Sambucus *nigra*, f.
    Nettle, 139. Urtica *dioica*, F.

April

25. Bindweed, *small*, 275.2. Convolvulus *arvens*. E.
Fir, *balm of Gilead*, Pinus *balsamea*, L.
Cicely, *wild*, 207.1. Chærophyllum *sylvestre*, F.
Young currants and gooseberries.

26. Plantain *ribwort*, 314.5. Plantago *lanceol.* F.
Germander, *wild*, 281.11. Veronica *chamæd.* F.
Cuckow pint, 266. Arum *maculatum, spatha out.*
Holly, 466. Ilex *aquifolium*, F.
Harebells, 373.3. Hyacinthus *nonscript.* F.

27. LILAC, 1763. H. Syringa *vulgaris*, F.
Crane's bill. *field*, 357.2. Geranium *cicutar.* F.
St. John's wort, 342.1. Hypericum *perforat.* E.
Betony *water*, 283.1. Scrophularia *aquat.* E.
Bryony, *white*, 261. Bryonia *alba*, E.
Birch tree, 443.1. Betula *alba*, F.

28. Jessamine, 1599.1. H. Jasminum *officinale*, l.
Thorn, *white*, 453.3. Cratægus *oxyacantha*, f.
*BLACK CAP, 79.12. *Motacilla atracapilla, sings.*

---

* The black cap is a very fine singing bird, and is by some in Norfolk called the mock nightingale. Whether it be a bird of passage i cannot say.

WHITE

April
28. * WHITE THROAT, 77. Motacilla
sylvia,
Juniper, 444.1. Juniperus *communis*, f.
Rasberry bush, 467.4. Rubus *idæus*, f.
Quince tree, 1452. H. Malus *Cydon*. f.
Crowfoot, *sweet wood*, 248.1. Ranunculus
auric. F.
29. Bugle, 245. Ajuga *reptans*, F.
Bay, 1688. H. Laurus *nobilis*, f.
Peas and beans, f.
Snow.
Chervil, *wild*, 207.1. Chærophyllum *temulent*. f.
Parsnep, *cow*, 205.1. Heracleum *sphondyl*. f.
Pine, *manured*, 1398.1. H. Pinus *pinea*. f.
30. *Snow.*
† *Thermom.* 5. *The lowest this month.*

* I have some doubt whether this bird be the Sylvia of Linnæus, though the description seems to answer to Ray's, and to one of my own, which I find among my papers.

† Vernal heat, according to Dr. Hales, at a medium, is 18.25.

V. MONTH.

All that is sweet to smell, all that can charm
Or eye or ear, bursts forth on every side,
And crouds upon the senses.

May 1. Crosswort, 223.1. Valantia *cruciata*, F.
Avens, 253.1. Geum *urbanum*, F.
Mugwort, 191.1. Artemisia *campestris*, E.
Bay, 1688. H. Laurus *nobilis*, L.

Lilly

May 3. Lilly of the valley, 264. Convallaria *Maialis*, f.

Violet, *water*, 285. Hottonia *palustris*, F.

4. Lettuce *lambs*, 201. Valeriana *locusta*, F.

Tulip tree, Liriodendron *tulipifera*, L.

Hound's tongue, 226.1. Cynoglossum *officinale*.

Cowslips, 284.3. Primula *veris*, F.

Valerian, *great wild*, 200.1. Valerian *officinalis*, F.

Rattle, *yellow*, 284.1. Rhinanthus *crista galli*, F.

Ice.

Thermom. 8. *The lowest this month.*

Fir, *silver, buds hurt by the frost.*

5. Twayblade, 385. Ophrys *ovata*, f.

Tormentil, 257. Tormentilla *erecta*, F.

Celandine, 309. Chelidonium *majus*, E.

Betony, 238.1. Betonica *officinalis*, E.

6. Oak, 440. Quercus, *robur*, F. et L.

Time for sowing barley.

Saxifrage, *white*, 354.6. Saxifraga *granulata*, F.

Ash, 469. Fraxinus *excelsior*, f.

Ramsons, 370.5. Allium *ursinum*, F.

Nettle, *white*, 240.1. Lamium *album*, F.

Quicken tree, 452.2. Sorbus *aucuparia*, F.

7. Fir, *Scotch*, 442. Pinus *sylvestris*, F.

8. Woodruffe, 224. Asperula *odorata*, F.

9. Chesnut tree, 1382. H. Fagus *castanea*, f.

10. Celandine, 309. Chelidonium *majus*, F.

Solomon's seal, 664. Convallaria *polygonat.* F.

Thorn, *white*, 453.3. Cratægus *oxyacantha*, F.

THE CALENDAR OF FLORA.

May
11. Maple, 470.2. Acer *campestre*, F.
    Roses, *garden*, f.
12. Barberry bush, 465. Berberis *vulgaris*, F.
    Chesnut, *horse*, 1683. H. Æsculus *hippocas*, F.
    Bugloss, *small wild*, 227.1. Lycopsis *arvensis*, F.
13. Grass, *water scorpion*, 229.4. Myosotis *scorpioid*, F.
    Quince tree, 1452. H. Pyrus *Cydonia*, F.
    Cleavers, 225. Galium *aparine*, F.
14. Mulberry tree, 1429. H. Morus *nigra*, L.
    Asp, 446.3. Populus *tremula*, l.
    Crowfoot, *bulbous*, 247.2. Ranunculus *bulbos*. F.
    Butter cups, 247. Ranunculus *repens*, F.
15. *Young turkies.*
    Lime tree, 473. Tilia *Europæa*, f.
    Milkwort, *287.1,2. Polygala *vulgaris*, F.
    Crane's bill, 359.10. Geranium *molle*, F.
    Walnut, 1376. H. Juglans *regia*, F.
16. Mustard, *hedge*, 298.4. Erysimum *officinale*, F.
20. Bryony, *black*, 262.1. Tamus *communis*, F.
    Many oaks, and more ashes and beeches, still without leaf.
    Violet, *sweet*, 364.1. Viola *odora*, D.
    Stitchwort, 346. Stellaria *holostea*, D.
    Anemone, *wood*, 259,1. Anemone *nemorosa*, D.
    Cuckow flower, 299.20. Cardamine *pratensis*, D.
    Earth nut, 209. Bunium, bulbocast. F.
    Mulberry tree, 1429. H. Morus *nigra*, f.
    Night-

May
21. Nightshade, 265. Atropa *belladonna*, f.
    RYE, 288. Secale *hybernum, in ear.*
23. Pellitory *of the wall,* 158.1. Parietaria *officin.* F.
24. Bramble, 467. Rubus *fruticosus,* f.
25. Moneywort, 283.1. Lysimachia *nummul.* F.
    Columbines, 173.1. Aquilegia *vulgar.* F. *in the woods.*
26. Tansy, *wild,* 256.5. Potentilla *anserina,* F.
    Henbane, 274. Hyoscyamus *niger,* F.
27. Campion, *white,* 339.8. Lychnis *dioica,* F.
    Clover, 328.6. Trifolium *pratense,* F.
28. Avens, 262.1. Geum *urbanum,* F.
    Chervil, *wild,* 207. Chærophyllum *temulent,* F.
30. Bryony, *black,* 262.1. Tamus *communis,* F.
    Brooklime, 280.8. Veronica *beccabunga,* F.
    Cuckow flower, 338. Lychnis *flos cuculi,* F.
    Cresses, *water,* 300.1. Sisymbrium *nasturt.* F.
    Thermom. 32. *Highest this month.*
31. Spurrey, 351.7. Spergula *arvensis,* F.
    Alder, *berry bearing,* 465. Rhamnus *frangula,* F.

VI. MONTH.

Now the mower whets his scythe,
And every shepherd tells his tale
Under the hawthorn in the dale. MILTON.

June 2. Elder, *water,* 460.1. Viburnum *opulus,* F.
    Lilly, *yellow water,* 368.1. Nymphæa *lutea,* F

THE CALENDAR OF FLORA.

June 2. Flower de luce, *yellow water*, 374. Iris *pseudo-acor*. F.
Mayweed, *stinking*, 185.3. Anthemis *cotula*, F.
Pimpernel, 282.1. Anagallis arvensis, F.
3. Arsmart, 145.4. Polygonum *persicaria*, F.
* Thyme, 430.1. Thymus *serpyllum*, F.
Parsnep, *cow*, 205. Heracleum *sphondylium*, F.
Quicken tree 452. Sorbus *aucuparia*, D.
5. Radish, *horse*, 301.1. Cochlearia *armorac*. F.
Thorn, *evergreen*, 1459.3. H. Mespilus *pyracantha*, F.
Bramble, 467. Rubus *fruticosus*, F.
† GOAT SUCKER, or FERN OWL, 27. Caprimulgus *Europæus*, is heard in the evening.
6. Vine, 1613. H. Vitis *vinifera*, b.
Flix weed, 298.3. Sisymbrium *sophia*, F.
Rasberry bush, 467.4. Rubus *idæus*, F.
Mallow, *dwarf*, 251.2. Malva *rotundifolia*, F.
Elder, 461.1. Sambucus *nigra*, F.
Stitchwort, *lesser*, 346. Stellaria *graminea* F.
Tare, *everlasting*, 320.3. Lathyrus *pratensis*, F.

---

* Pliny, lib. 11. §. 14. says, the chief time for bees to make honey is about the solstice, when the vine and thyme are in blow. According to his account then these plants are as forward in England as in Italy.

† This bird is said by Catesby, as quoted by the author of the treatise De Migrationibus Avium, to be a bird of passage.

June 6. Gout *weed*, 208.3. Ægopodium *podagrar.* F.

Bryony, *white*, 261.1,2. Bryonia *alba*, F.
Rose, dog, 454.1. Rosa *canina*, F.
Bugloss *vipers*, 227.1. Echium *vulgare*, F.

7. Grass, *vernal*, 398.1. Anthoxanthum *odorat.* F.
Darnel, *red*, 395. Lolium *perenne*, F.
Poppy, *wild*, 308.1. Papaver *somnifer*, F.
Buckwheat, 181. H. Polygonum *fagopyrum*, F.

8. Pondweed, *narrow leaved*, 145.9. Polygonum *amphib.* F.
Sanicle, 221.1. Sanicula *Europæa*, F.

9. Eyebright, *284.1. Euphrasia *officinalis*, F.
Heath, *fine leaved*, 471.3. Erica *cinerea*, F.
Saxifrage, *bugle, hyacinth*, D.
Broom, 474.1. Spartium *scoparium, podded.*
Nettle, *hedge*, 237. Stachys *sylvatica*, F.

12. Wheat, 386.1. Triticum *hybernum, in ear.*
Meadow sweet, 259.1. Spiræa *ulmaria*, f.
Scabious, field, 191.1. Scabiosa *arvensis*, F.
Valerian, *great water*, 200.1. Valeriana *officinal.* f.
Cinquefoil, *marsh*, 256.1. Comarum *palustre*, F.
Orchis, *lesser butterfly*, 380.18. Orchis *bifolia*, F.

13. Willow herb, *great hairy*, 311.2. Epilobium *hirsutum*, F.
Parsnep, *cow*, 205. Heracleum *sphondyl.* F.
Betony, *water*, 283.1. Scrophularia *aquat.* F.

Cockle,

THE CALENDAR OF FLORA.

June
13. Cockle, 338.3. Agrostemma *githago*, F.
    Sage, 510.7. H. Salvia *officinalis*, F.
15. Mallow, 251.1. Malva *sylvestris*, F.
    Nipplewort, 173 1. Lapsana *communis*, F.
    Woodbind, 458.1, 2. Lonicera *periclymen.* f.
    NIGHTINGALE *sings.*
16. Fir, *Weymouth*, 8 dend. Pinus *tæda*, F.
    Hemlock, 215.1. Conium *maculatum*, F.
    Nightshade, *woody,* 265. Solanum *dulcamara*, F.
    Archangel, *white*, 240 Lamium *album*, F.
17. Vervain, 236. Verbena *officinalis*, F.
    Agrimony, 202. Agrimonia *eupator*, F.
    Hemlock, *water,* 215. Phellandrium *aquatic.* F.
    Acacia, 1719. H. Robinia *pseudo-acacia*, F.
18. Yarrow, 183. Achillea *millefolium*, F.
19. *Thermom.* 44.25. *Highest this month.*
21. Orache, *wild,* 154.1. Chenopodium *album,* F.
    Solstice. About this time ROOKS *come not to their nest trees at night.*
    Wheat, 386.1. Triticum *hybernum*, F.
    Rye, 388.1. Secale *hybernum*, F.
    Self-heal, 238. Prunella *vulgaris,* f.
    Parsley, *hedge,* 219.4. Tordylium *anthriscus,* f.
    Grasses of many kinds, cs festuca, aira, agrostis, phleum cynosurus, in ear.
22. Horehound, *base,* 239. Stachys *Germanica,* F.

X St.

June

22. St. John's wort, 342. Hypericum *perforatum*, F.
    Parsnep, 206.1. Pastinaca *sativa*, F.
    Mullein, *white*, 287. Verbascum *thapsus*, F.
    Poppy, *wild*, 308. Papaver *somnifer*, F.
23. Larkspur, 708.3. H. Delphinium *Ajacis*, F.
    Marygold, *corn*, 182.1. Chrysanthemum *seget*. F.
24. Rosemary, 515. H. Rosmarinus *officinalis*, D.
25. Vine, 1613. H. Vitis *vinifera*, F.
    Bindweed, *great*, 275.2. Convolvulus *arvensis*, F.
    Feverfew, 187. Matricaria *parthenium*, F.
    Woad, *wild*, 366.2. Reseda *luteola*, F.
    Rocket, *base*, 366.1. Reseda *lutea*, F.
    Archangel, *yellow*, 240.5. Galeopsis *galeobdolon*, F.
    Wheat, 386.1. Triticum *hybernum*, F.
    Thermom. 20. *The lowest this month.*
27. *Clover mowed.*
    Pennywort, *marsh*, 222. Hydrocotule *vulgaris*, F.
    Meadow, *sweet*, 259. Spiræa *ulmaria*, F.
28. Oats *manured*, 389. Avena *sativa*, F.
    Barley, 388. Hordeum *vulgare*, F.
    Midsummer shoots of apricot, oak, beech, elm.
    Succory, wild, 172.1. Cichorium *intybus*, F.
    Blue bottles, 198. Centaurea *cyanus*, F.

THE CALENDAR OF FLORA.

June
- 28. Knapweed, *great*, 198. Centaurea *scabiosa*, F.
- 30. *Currants ripe.*
  According to Dr. Hales, May and June heat is, at a medium, 28.5.

---

* The groves, the fields, the meadows, now no more
With melody resound. 'Tis silence all,
As if the lovely songsters, overwhelm'd
By bounteous nature's plenty, lay intranc'd
In drowsy lethargy.

---

* I heard no birds after the end of this month, except the STONE CURLEW, 108.4. Charadrius Oedicnemus, whistling late at night; the YELLOW HAMMER, 93.2. Emberiza flava; the GOLDFINCH, 89.1. and GOLDEN CRESTED WREN, 79.9. Motacilla regulus, now and then chirping. I omitted to note down when the cuckow left off singing, but, as well as I remember, it was about this time. Aristotle says, that this bird disappears about the rising of the dog star, i. e. towards the latter end of July.

VII. MONTH.

Berries and pulpous fruits of various kinds,
The promise of the blooming spring, now yield
Their rich and wholesome juices, meant t'allay
The ferment of the bilious blood.

July 2. Beech, 439. Fagus *sylvatica*, F.
  Pearlwort, 345.2. Fagina *procumbens*, F.
  Carrot, *wild*, 218. Daucus *carrota*, F.
  Grass, *dog*, 390.1. Triticum *repens, in car.*
  Violet, *Calathian*, 274. Gentiana *pneumonan*, F.
- 4. Silver weed, 256.5. Potentilla *anserina*, F.
  Betony, 238.1. Betonica *officinalis*, F.
  Nightshade, *enchanters*, 289. Circæa *lutetiana*, f.

July 6. Lavender, 512. Lavendula *spica*, F.
Parsley, *hedge*, Tordylium *anthriscus*, F.
Gromill, 228.1. Lithospermum *officinale*, F.
Furze, 473. Ulex *genista*, D.
Cow wheat, *eyebright*, 284.2. Euphrasia *odont*. F.

7. Pinks, maiden, 335.1. Dianthus *deltoides*, F.

8. Tansey, 188.1. Tanacetum *vulgare*, f.
Bed-straw, *lady's yellow*, 224. Galium *verum*, F.
Sage, *wood*, 245. Teucrium *scorodonia*, F.
Spinach, 162. H. Spinacia *oleracia*, F.
Thermom. 22. Lowest this month.

9. Angelica, *wild*, 208.2. Angelica *sylvestris*, F.
Strawberries ripe.
Fennel, 217. Anethum *fœniculum*, F.

10. Beans, *kidney*, 884. H. Phaseolus *vulgaris*, *podded*.
Parsley, 884. H. Apium *petroselinum*, F.
Sun dew, *round leaved*, 356.3. Drosera *rotundifol*. F.
Sun dew, *long leaved*, 356.4. Drosera *longifol*. F.
Lilly, *white*, 1109. H. Lilium *candidum*, f.

11. Mullein, *hoary*, 288. Verbascum *phlomoid*. F.
Plantain, *great*, 314.1,2. Plantago *major*, F.
WILLOW, SPIKED, of Theophr. 1699. H. Spiræa *salicifol*. F.
Jessamine, 1599. H. Jasminum *officinale*, F.
Rest harrow, 332. Ononis *spinosa*, F.

Hyssop,

THE CALENDAR OF FLORA.

July
11. Hyssop, 516. H. Hyssopus *officinalis*, F.
    Potatoes, 615.14. H. Solanum *tuberosum*, F.
    Second shoots of the maple.
    Bell flower, *round leaved*, 277.5. Campanula, F.
    LILLY, WHITE, 1109. H. Lilium *candidum*, F.
    Rasberries ripe.
    Figs yellow.
13. LIME TREE, 473. Tilia *Europæa*, F.
    Knapweed, 198.2. Centaurea *jacea*, F.
    Stonecrop, 269. Sedum *rupestre*, F.
    Grass, *knot*, 146. Polygonum *aviculare*, F.
    Grass, *bearded dog*, 390.2. Triticum *caninum*, F.
15. *Thermom. 39. Highest this month.*
16. Asparagus, 267.1. Asparagus *officinalis, berries.*
    Mugwort, 190.1. Artemisia *vulgaris*, F.
18. Willow herb, *purple spiked*, 367.1. Lythrum *salicaria*, F.
    YOUNG PARTRIDGES.
    Agrimony, *water hemp*, 187.1. Bidens *tripart.* F.
20. Flax, *purging*, 362.6. Linum *catharticum*, F.
    Arsmart, *spotted*, 145.4. Polygonum *persicaria*, F.
    Lilly, *martagon*, 1112. H. Lilium *martagon.*
    HENS *moult.*
22. Orpine, 269. Sedum *telephium*, f.
    Hart's tongue, 116. Asplenium *scolopendra*, F.

Penny

July

22. Pennyroyal, 235. Mentha *pulegium*, F.
   Bramble, 461.1. Rubus *fruticosus*. *Fruit red.*
   Laurustinus, 1690. H. Viburnum *tinus*, f.
24. Elecampane, 176. Inula *helenium*, F.
   Amaranth, 202. H. Amaranthus *caudatus*, F.
27. Bindweed, *great*, 275.1. Convolvulus *sepium*, F.
28. Plantain, *great water*, 257.1. Alisma *plantago*, F.
   Mint, *water*, 233.6. Mentha *aquatica*, F.
   Willow herb, 311.6. Epilobium *palustre*, F.
   Thistle tree sow, 163.7. Sonchus *arvensis*, F.
   Burdock, 197.2. Arctium *lappa*, f.
   Saxifrage, *burnet*, 213.1,2. Pimpinella, *saxifraga*, F.
   DEVIL'S BIT, 191.3. Scabiosa *succisa*, F.
32. Nightshade, common, 288.4. Solanum *nigrum*, F.
   DOVE, RING, 62.9. Columba *palumbus*, cooes.

VIII. MONTH.

Pour'd from the villages, a numerous train
Now spreads o'er all the fields. In form'd array
The reapers move, nor shrink for heat or toil,
By emulation urg'd. Others dispers'd,
Or bind in sheaves, or load or guide the wain
That tinkles as it passes. Far behind,
Old age and infancy with careful hand
Pick up each straggling ear, &c.

THE CALENDAR OF FLORA.

August
1. Melilot, 331.1. Trifolium *officinale*, F.
Rue, 874.1. Ruta *graveolens*, F.
Soapwort, 339.6. Saponaria *officinalis*, F.
Bedstraw, *white, lady's*, 224.2. Galium *palustre*, F.
Parsnep, *water*, 300. Sisymbrium *nasturt*. F.
Oats almost fit to cut.
3. *Barley cut.*
5. Tansy, 188.1. Tanacetum *vulgare*, F.
Onion, 1115. H. Allium *cepa*, F.
7. Horehound, 239. Marrubium *vulgare*, F.
Mint, *water*, 233.6. Mentha *aquat*. F.
Nettle, 139. Urtica *dioica*, F.
Orpine, 269.1. Sedum *telephium*, F.
NUTHATCH, 47. Sitta *Europæa*, *chatters*.
8. *Thermom.* 20. *Lowest to the 27th of this month*
9. Mint, *red*, 232.5. Mentha *gentilis*, F.
Wormwood, 188.1. Artemisia *absinthium*, F.
12. Horehound, *water*, 236.1. Lycopus *Europæus*, F.
Thistle, *lady's*, 195.12. Carduus *marianus*, F.
Burdock, 196. Arctium *lappa*, F.
ROOKS *come to the nest trees in the evening, but do not roost there.*
14. Clary, *wild*, 237.1. Salvia *verbenaca*, F.
STONE CURLEW, 108. Charadrius *oedicnemus whistles at night.*
15. Mallow, *vervain*, 252. Malva *alcea*, F.

THE CALENDAR OF FLORA,

August
- 15. GOAT SUCKER, 26.1. Caprimulgus *Europæus, makes a noise in the evening, and young owls.*
- 16. *Thermom. 35. The highest to the 27th of this month.*
- 17. Orach, *wild*, 154.1. Chenopodium *album*.
  ROOKS *roost on their nest trees.*
  GOAT SUCKER, *no longer heard.*
- 21. *Peas and wheat cut.*
  Devil's bit, *yellow*, 164.1. Leontodon, *autumnal*. F.
- 26. ROBIN RED BREAST, 78.3. Motacilla *rubecula, sings*.
  Goule, 443. Myrica *gale*, F. R.
  Golden rod, *marsh*, 176.2. Senecio *paludosus*, F.
- 29. Smallage, 214. Apium *graveolens*, F.
  Teasel, 192.2. Dipsacus *fullonum*, F.
  Vipers come out of their holes still.

IX. MONTH.

How sweetly nature strikes the ravish'd eye
Thro' the fine veil with which she oft conceals
Her charms in part, as conscious of decay!

September
- 2. WILLOW HERB, *yellow*, 282.1. Lysimachia *vulgaris*, F.
  Traveller's joy, 258. Clematis *vitalba*, F.

---

* From the 27th of this month to the 10th of September I was from home, and therefore cannot be sure that I saw the first blow of the plants during that interval.

THE CALENDAR OF FLORA.

September
- 5. Grafs of Parnaſſus, 355. Parnaſſia *paluſtris.*
- 10. *Catkins of the haſel formed.*
  Thermom. 17. *The loweſt from the 10th to the end of this month.*
- 11. *Catkins of the birch formed.*
  Leaves of the Scotch fir fall.
  Bramble ſtill in blow, though ſome of the fruit has been ripe ſome time ; ſo that there are green, red, and black berries on the ſame individual plant at the ſame time.
  Ivy, 459. Hedera *helix,* f.
- 14. *Leaves of the ſycomore, birch, lime, mountain aſh, elm, begin to change.*
- 16. Furze, 475. Ulex *Europæus,* F.
  Catkins of the alder formed.
  Thermom. 36.75. *The higheſt from the 10th to the end of this month.*
  CHAFFINCH, 88. Fringilla *cælebs, chirps.*
- 17. *Herrings.*
- 20. FERN, FEMALE, 124.1, Pteris *aquilina,* turned brown.
  Aſh, *mountain,* 452.2. Sorbus *aucuparia,* F. R.
  Laurel, 1549. H. Prunus *laurocerasus,* f. r.
  Hops, humulus *lupulus,* 137.1. f. r.
- 21. SWALLOWS *gone. Full moon.*
- 23. *Autumnal æquinox.*
- 25. WOOD LARK, 69.2. Alauda *arborea, ſings.*
  FIELD FARE, 64.3. Turdus *pilaris, appears,*

Leaves

September.

25. *Leaves of the plane tree, tawny—of the hasel, yellow—of the oak, yellowish green—of the sycomore, dirty brown—of the maple, pale yellow—of the ash, fine lemon—of the elm, orange—of the hawthorn, tawny yellow—of the cherry, red—of the hornbeam, bright yellow—of the willow, still hoary.*
27. BLACK BIRD *sings.*
29. THRUSH, 64.2. Turdus *musicus, sings.*
30. *Bramble, 467.1. Rubus *fruticosus*, F.

*Autumnal heat, according to Dr. Hales, at a medium, is 18. 25.

X. MONTH.

Arise, ye winds, 'tis now your time to blow,
And aid the work of nature. On your wings
The pregnant seeds convey'd shall plant a race
Far from their native soil.

October

1. Bryony, *black*, 262. Tamus *communis*, F.R.
Elder, *marsh*, 460.1. Viburnum *opulus*, F.R.
Elder, 461.1. Sambucus *nigra*, F. R.
Briar, 454.1. Rosa *canina*, F. R.
Alder, *black*, 465. Rhamnus *frangula*, F.R.
Holly, 466. Ilex *aquifolium*, F. R.
Barberry, 465. Berberis *vulgaris*, F. R.
Nightshade, *woody*, 265. Solanum *dulcamara*, F. R.
2. Thorn, *black*, 462.1. Prunus *spinosa*, F. R.
CROW,

THE CALENDAR OF FLORA.

October
2. *CROW, ROYSTON, 39.4. Corvus *cornix*, returns.
5. *Catkins of sallows formed.*
6. *Leaves of asp almost all off—of chesnut, yellow—of birch, gold-coloured.*
Thermom. 26.50. Highest this month.
7. BLACK BIRD, 65.1. Turdus *merula*, sings.
Wind high; rooks sport and dash about as in play, and repair their nests.
9. *Spindle tree, 468.1. Euvonymus* Eu*ropæus*, F. R.
Some ash trees quite stripped of their leaves. Leaves of marsh elder of a beautiful red, or rather pink colour.
10. WOOD LARK *sings.*
§ RING DOVE *cooes.*
14. WOOD LARK *sings.*
Several plants still in flower, as pansy, white hehn, black nonesuch, hawkweed, bugloss, gentian, small stitchwort, &c. in grounds not broken up.
A great mist and perfect calm; not so much as a leaf falls. Spiders webs innumerable appear every where. Woodlark sings. Rooks do not stir but sit quietly on their nest trees.
16. GEESE, WILD, 136.4. Anas, *anser*, *leave the fens and go to the rye lands.*

---

* Linnæus observes in the Systema Naturæ, and the Fauna Suecica, that this bird is useful to the husbandman, tho' ill treated by him.

§ Aristotle says, this bird does not cooe in the winter, unless the weather happens to be mild.

October

    22. WOODCOCK, 104. Scolopax *rusticola*, *returns*.

     Some *ash-trees still green*.

    24. LARK, SKY, 69.1. Alauda *arvensis, sings*.

     Privet, 465.1. Ligustrum *vulgare*, F. R.

    26. *Thermom.* 7. *Lowest this month.*

     Honeysuckle, 458.1,2. Lonicera *periclymen. still in flower in the hedges, and mallow and feverfew.*

     WILD GEESE *continue going to the rye lands*.

Now from the north
Of Norumbega and the Samoeïd shore,
Bursting their brazen dungeons, arm'd with ice,
And snow, and hail, and stormy gust, and flaw,
Boreas, and Cæcias, and Argestes loud,
And Thrascias rend the woods, and seas up-turn.
                      MILTON.

    Here ends the Calendar, being interrupted by my going to London. During the whole time it was kept, the barometer fluctuated between 29.1. and 29.9. except a few days, when it sunk to 28.6. and rose to 30 ¼.

A Sibirian *or* Lapland Year.

June
    23. Snow melts.
July 1. Snow gone.
    9. Fields quite green.
   17. Plants at full growth.
   25. Plants in full blow.
August
    2. Fruits ripe.
   10. Plants shed their seeds.
   18. Snow.
      From this time to June 23, snow and ice; so that by this account, plants, from the coming out of the ground to the ripening of their seeds, take but a month. And the spring, summer and autumn, are crouded into the space of 56 days. This account is taken from a treatise published in the Amæn. Academ. vol. iv. and agrees with one i have seen quoted out of Gmelin, who was in Sibiria many years.

THE

CALENDAR of FLORA.

By THEOPHRASTUS.

At ATHENS, Latitude 37° 25.

INTRODUCTION.

THE following Calendar was extracted chiefly from Theophrastus's History of plants, and put together in the best manner i was able from imperfect materials. Any one who looks into the original, will see that accuracy ought not to be expected; the manner of marking the times being often very indeterminate.

I am sensible that objections may be made to many parts of this Calendar, but i thought it not worth while to give my reasons for what I have done, and thereby load a piece of mere curiosity with pompous quotations.

It has always seemed extraordinary to me, that when disciples of Linnæus have been sent into so many parts of the world, in order to make discoveries in natural history, viz. Asia, Pensylvania, Lapland, Ægypt, Palestine, Malabar, Surat, China, Java, Spain, America, Gotland, Italy, Apulia, Surinam, and St. Eustatia, that Greece should have been overlooked. It is true, Monsieur Tournefort was sent into the Levant by Lewis the Fourteenth to search for plants, and spent some years there; it is also as true, that he had all the knowledge and zeal necessary for such a commission; but the country was too extensive for one man to examine thoroughly in that space of time. He rambled over most of the Greek islands, Armenia, and other parts of Asia; and though he enriched the royal gardens with many new plants, yet several must have escaped him for want of time, or a proper season.

It were to be wished, therefore, that some persons properly qualified, might be sent to Greece, and be enjoined to make Attica, particularly, their place of residence for a year at least. This might furnish a Flora and Fauna Attica, that would

would be extremely curious to all lovers of natural history; and tend to clear up many passages in those authors, who first opened that branch of knowledge, as well as carried some parts of it much farther, than is generally known, or at least acknowledged; and from whose writings much more benefit might still be reaped, were they better understood, especially in the medicinal way.

As the English nation will have the honour of first making known to the world the true and accurate proportions of the ancient Greek architecture, so i hope it is reserved for us to bring the rest of Europe thoroughly acquainted with the nature of the soil, climate, productions, animals, &c. of a country whose ancient glory so much resembles our own, and in a great measure has been the cause of it, by furnishing us with the best models of good sense, taste, and just sentiments in every branch of human knowledge: We therefore ought in a particular manner to look upon Attica, from whence, as Cicero says, *Humanitas, doctrina, fruges, jura, leges ortæ, atque in omnes terras distributa putantur*, with the veneration due to a mother country. Should such a scheme take place, i could name a person, perfectly well qualified, by his youth and abilities, and zealously inclined upon proper encouragement to be one of the party. France, Sweden*, and Russia have set us examples of this kind, and why this great and flourishing nation should not follow them, i cannot see. We have had our share in advancing natural history, it is true, but hitherto without any public encouragement.

---

* Amongst many instances of this sort, there is one that deserves particular notice mentioned Amæn. Academ. p. 445. the author says, that Hasselquist was sent into Ægypt at the expence of his countrymen the East Gothlanders, of the heads of the university, and of the East India company, for the study of natural history; and staid above a year at Cairo.

THE
CALENDAR of FLORA.
By THEOPHRASTUS.

Feb.
1. —* Violet, *early bulbous*, 1144. H. Leucoium *vernum*, λευκοιον, F.
    Wall flower, 291.2. Cheiranthus, *cheiri*, φλογιον, F.
    Cornel tree, 1536. H. Cornus, *mas*, κρανεια, L.
    Dogberry, 460. Cornus, *sanguinea*, θηλυ κρατεια, L.
14. — Bay tree, 1688. H. Laurus *nobilis*, δαφνη, L.
    Alder, 442. Betula *alnus*, κληθρα, L.
    Abele, 446.3. Populus *alba*, λευκη, L.
    Elm, 468. Ulmus *campestris*, πτελεα, L.
    Sallow, Salix, ιτεα, L.
    Poplar, *black*, 446.1. Populus *nigra*, αιγειρος, L.
    Plane tree, 1706. H. Platanus *orient.* πλατανος, L.

---

* This mark — after some of the figures, denotes that the time is only determined within certain limits.
All the other marks mean the same as in my own Calendar.

THE CALENDAR OF FLORA.

March
12 — * Beginning of SPRING.
 Fig tree, 1431. H. Ficus *carica*, ερινεος, L.
 Alaternus, 1608.1.   Rhamnus *alatern.* φιλυκη, L.
 Hawthorn, 453 3.   Cratægus *oxyacanth.* οξυακανθος, L.
 Christ's thorn, 1708. H. Rhamnus *paliurus*, παλιηρος, L.
 Turpentine tree, 1577. H. Pistacia *terebin.* τερμινθος, L.
 Chesnut-tree, 1382. H. Fagus *castanea*, διος βαλανος, L.
 Walnut-tree, 1376. H. Juglans *regia*, καρυα, L.
 Lilly of the valley, 264.   Convallaria *Maialis*, οιανθη, F.
 Narcissus, C. B. 49. ανεμωνη λειμωνια, F.
 Daffodil, 1131. H. Narcissus *pseudo narc.* βυλβοκωδιον, F.
 Corn flag, 1169.2. H. Gladiolus *communis*, F.
 Hyacinth, 1162.31. Hyacinthus *comosus*, υακινθος, F.
 Rose, rosa, ροδον, F.
20. † Elder tree, 461.1. Sambucus *nigra*, ακτη, L.

---

* Between February 28 and March 12, the Ornithian winds blow, and SWALLOW appears.
† Between March 11 and 26, the *kite* and *nightingale* appear, that is in the leafing season. The appearance of the hawk is consonant to what Aristotle says, as quoted in the preface, but is determined upon a different kind of testimony; which is a proof that this part of the Calendar at least is tolerably well stated.

THE CALENDAR OF FLORA.

March
20. Fleawort, 881. H. Plantago *psyllium*, κυνοψ, F.
    Oak, 442. Quercus, robur, δρυς, L.
    Fig-tree, 1431. H. Ficus *carica*, συκη, L.
    Oak, 1386. H. Quercus, *esculus*, φηγος, L.
    Lime-tree, 473. Tilia *Europæa*, φιλυρα, L.
    Maple, 470.2. Acer *campestris*, ζυγια, L.
    Apple-tree, 451. Pyrus *malus*, μηλεα, L.
    Ivy, 459. Hedera *helix*, ιψος, L.
    Beam tree, *white*, 453. Cratægus *aria*, αρια, L.
26. Tree of life, 1408. H. Thuia *occident*. θυεια, L.

April
4. — Succory, 172. Cichorium *intybus*, κιχορειον, F.

May 12.   Beginning of SUMMER.
15 — *Wheat harvest.*
    Turpentine tree, 1557. H. Pistacia *terebin*, τερμινθος, F. R.
    Flower of Constantinople, 992.1. H. Lychnis *Chalced.* λυχνις, F.
    Rose campion, 993.2. H. Lychnis *coronar.* διος ανθος, F.
    Asphodel, *yellow*, 1192.4. H. Asphodelus *luteus*, αμαρακος, F.
    Ash-tree, 468. Fraxinus *excelsior*, μελια, F. R.
    Maple, 470.2. Acer *pseudo-platanus*, σφενδαμνος, F. R.
    Pine, 1398. H. Pinus *sylvestris*, πιτυς, F.
    Fir-tree, *common*, 1396.2. Pinus *abies*, πευκη, F.

THE CALENDAR OF FLORA.

June
20 — * Fir-tree, *yew leaved*, 1394. Pinus *picea*, ελατη, F.
  Yew-tree, 445. Taxus *baccata*, μιλθ, F. R.
  Cornel tree, 1536. H. Cornus *mas*, κρανεια, F. R.
  Midsummer shoots of the oak.
  The fig, the vine, and the pomegranate, shoot later.

July
23. Cuckow disappears.
30. Etesian winds blow.

---

* Botanists doubt which of these two firs is the πευκη and which the ελαιη. Theophrastus says expressly that the πευκη flowers some days before the ελαιη, and therefore this question might be probably decided.

August
19 —   Beginning of A U T U M N.
  Lilly, Lilium, λειριον, F.
  Crocus, 1173.3. Crocus *autumnal*. κροκθ, F.
  Dogberry 460. Cornus *sanguinea* θηλυκρανεια, F. R.
  Alder, 442. Betula *alnus*, κληθρα, F. R.
  Quail, 58.6. Tetrao, *coturnix*, ορτυξ, departs.

Sept.
20 — Crane, 95. Ardea, *grus*, γερανος, departs.
  Autumn shoots of trees.

October
12 — Oak, 440. Quercus, *robur*, δρυς, F. R.
                                            Chesnu

THE CALENDAR OF FLORA.

October
12 — Chesnut, 1382. H. Fagus, *castanea*, διος βαλανος, F. R.
Christ's thorn, 1708. H. Rhamnus *paliur.* παλιυρος, F. R.
Hawthorn, 453.3. Cratægus *oxyacantha*, οξυακανθος, F. R.
Holm oak, 1391. H. Quercus *coccifer*, πρινος, F. R.
Alaternus, 1608.1. H. Rhamnus *alatern.* φιλυκη, F. R.

29 — Venice sumach, 1696. H. Rhus *cotinus*, κοκκομηλεα, F.
Apple-tree, 451. Pyrus *malus*, μηλεα, F. R.
Beam-tree, *white*, 453. Cratægus *aria*, αρια, F. R.
Lime-tree, 473. Tilia *Europæa*, φιλυρη, F. R.
Box-tree, 445. Buxus *sempervivens*, πυξος, F. R.

Beginning of WINTER.

Novem.
15 — Ivy, 459. Hedera *helix*, κιτθος, F. R.
Juniper, 444. H. Juniperus *communis*, αρκευθος, F. R.
Tree of life, 1408. H. Thuia *occident.* θυεια, F. R.
Yew-tree, 445. Taxus *baccata*, μιλος F. R.
Pear-tree, 1450. Pyrus *communis*, αχρας, F. R.
Arbutus, 1577. 2. H. αδραχνη, F. R.

INDEX.

☞ The large Roman Numerals refer to the Months of both Calendars; the small Numerals to the Days of the Month of the Swedish; the common Figures to the English.

A.

ACER, IV. 6. 13. 18. 25. V. 2.
    Achillea, VI. xxiii. xxix. VII. vi. vii. VI. 18.
Actæa, VI. vii.
Adonis, VI. xiii. ix. i.
Adoxa, V. viii.
Aegopodium, VI. 6.
Aesculus, V. xvi. IV. 1. V. 12.
Agrimonia, VII. vi. IV. 24. V. 17.
Agrostema, VII. v. VI. 13.
Ajuga, V. xxx. IV. 29.
Alauda, III. xx. II. 4. IX. 25. X. 10. 14. 24.
Alchemilla, V. xxv.
Alisma, VII. 28.
Allium, V. 6. VIII. 5.
Allopeurus, VI. iv.
Alnus, III. 26.
Alsine, I. 26.
Amaranthus, VII. 24.
Amygdalus, IV. 6.
Anagallis, VI. 2.
Anas, IV. vii. II. 12. X. 16. 26.
Anemone, IV. xvi. V. iii. IV. 10. V. 20.
Anethum, VII. 9.
Angelica, VII. 9.
Anthemis, VI. xxiii. xxix. VI. 2.
Anthericum, VI. ii.
Anthyllis, VI. xii.
Anthoxanthum, VI. 7.
Antirrhinum, VI. xxviii.
Apis, III. 2. IV. 10.
Apium, VII. 10. VIII. 29.
Aquifolium, IV. 3.
Aquilegia, VI. vii. V. 25.
Arctium, VII. ix. VII. 28.
Ardea, IV. 14.
Artemisia, VII. x. VIII. xiv. IX. vi. IV. 9. V. 1. VII. 16. VIII. 9.
Arum, IV. 26.
Asparagus, VII. 16.

Asperula, V. 8.
Atriplex, IX. xxii.
Atropa, VI. xviii. IV. 18. V. 21.
Asarum, V. iii.
Astragalus, VI. xxiv.
Avena, VI. 28. VIII. 2.

B.

Ballota, VII. x.
Bellis, I. 26.
Berberis, V. xiii. VI. viii. V. 12. X. 1.
Betonica, V. v.
Betula, V. xiii. xiv. IV. 1. 7. 27.
Bidens, VII. 18.
Brassica, IV. 15.
Briza, VI. xviii.
Bryonia, VI. xxviii. IV. 27. VI. 6.
Bunium, V. 20.

C.

Caltha, V. xxi. IV. 16.
Campanula, VI. xi. xvi. xxix. VII. i. ii. vi. VII. 11.
Cannabis, VII. ix.
Caprimulgus, V. 5. VIII. 15.
Cardamine, V. xxviii. IV. 18. V. 20.
Carduus, VI. xxii. VII. ix. xvii. VIII. 12.
Carpinus, V. xvi. III. 29. IV. 10. 13. 25.
Castanea, IV. 16.
Centaurea, VI. xxii. xxiv. VII. i. vi. xxii. VI. 28. VII. 13.
Chærophillum, V. xxv. VII. vii. IV. 25. 29. V. 28. VI. 5.
Charadrius, VIII. 14.
Cheiranthus, IV. 21.
Chelidonium, V. xxvii. V. 5. 10.
Chenopodium, IV. 24. VI. 21. VIII. 17.
Chrysanthemum, VI. xvii. xxix. VI. 23.
Chrysoplenium, V. iii.

                                                        Cichorium

INDEX.

Cichorium. VII. xii. VI. 28.
Ciconia, V. ix.
Circæa, VII. x. VII. iv.
Cistus, VI. xx.
Clematis, IX. vi. IX. 2.
Clinopodium, VII. x.
Cochlearia, III. 21. VI. 5.
Colchicum, VII. xviii. VIII. xxviii.
Columba, III. 5. VII. 30. X. 10.
Comarum, VI. xx. VI. 12.
Conium, VI. 16.
Convallaria, V. xxx. V. 3, 10.
Convolvulus, VII. i. iv. IV. 25. VII. 27.
Conyza, VII. xxiii.
Cornus, V. xxx.
Corvus, II. 12. III. 2. VIII. 27. X. 2.
Corylus, IV. xii. V. ix. I. 23. II. 22. IV. 7.
Cratægus, V. xv. VI. xii. xvii. III. 29. IV. 28. V. 10.
Crepis, VI. xxix.
Crocus, IV. xiii.
Cucubalus, VIII. i.
Cuculus, V. xii. IV. 17.
Cygnus, IV. x.
Cynoglossum, IV. 10. V. 4.
Cyprinus, V. xxv. xxx.
Cytisus, IV. xiii.

D.
Dactylis, VII. xxii.
Daphne, IV. xiii. IV. 16.
Daucus, VII. vi. VII. 2.
Delphinium, VI. 23.
Dianthus, VI. xxiv. xxix. VII. viii. VII. 7.
Dies Chalybeati, II. xxii.
Digitalis, VII. iv.
Dipsacus, VII. xviii. xx. VIII. 29.
Draba, IV. xv.
Drosera, VII. x.

E.
Echium, VI. 6.
Elymus, VI. xxviii.
Empetrum, IV. xxx.
Epilobium, VI. xxiv. xxvi. VII. xii. xv. VI. 13. VII. 28.
Erica, VI. xxix. VI. 9.
Erigeron, VII. xv.

Eriophorum, V. xxx.
Erysimum, V. xxv. IV. 10. 16. V. 16.
Esox, IV. x.
Euonymus, V. xiv. X. 9.
Eupatorium, VII. xv.
Euphorbia, IV. 25.
Euphrasia, VI. xvii. VI. 9. VII. 6.

F.
Fagus, V. xvi. IV. 18. 21. V. 9. VII. 2.
Ficus, IX. iv. IV. 16. 21. 22. VII. 11.
Filipendula, VI. xxii.
Fragaria, VI. xxvi. IV. 13. VII. 9.
Frankenia, VII. ix.
Fraxinus, V. xxi. IV. 18. 22. V. 6.
Fringilla, II. 16. IX. 16.

G.
Galanthus, IV. xiii. I. 26.
Galeopsis, VI. 25.
Galium, VI. xvi. VII. vi. IV. 3. V. 13. VII. 8. VIII. 1.
Genista, IV. 22.
Gentiana, VI. xxii. VII. 26. VII. 2.
Geranium, V. xxvi. xxx. VI. v. ix. xi. xiv. IV. 23. 27. V. 15.
Geum, VI. ii. xiii. V. 1. 28.
Glechoma, V. xxvi. IV. 16.
Gnaphalium, V. xxvi.

H.
Hedera, VIII. 9. IX. 11.
Helenium, VII. 24.
Helleborus, IV. xxi.
Hemerocallis, VI. xx. VII. vii. xvi.
Heracleum, VII. i. III. 28. IV. 29. VI. 3. 13.
Herniaria, VI. xxix.
Hesperis, VI. vii.
Hieracium, VII. xvi.
Hippophae, V. xiv.
Hirundo, V. ix. IV. 6. IX. 21.
Holcus, VII. v.
Hordeum, V. xiii. V. 6. VIII. 3.
Hottonia, V. 3.
Humulus, VII. vi. IV. 14. IX. 20.
Hyacinthus, IV. 26.
Hybernacula, V. viii.
Hydrocotule, VI. xxvii.
Hyoscyamus, VI. xii. V. 26.
Hypericum,

INDEX.

Hypericum, VI. xxix. VII. vi. xii. IV. 27. VI. 21.
Hypochæris, VI. xxvii.
Hyssopus, VII. 11.

I.

Jasminus, IV. 28. VII. 11.
Ilex, I. 23. IV. 24. 26. X. 1.
Impatiens, VII. xviii.
Inula, VII. xiii. xxiv. IX. xvii.
Iris, VI. x. VII. vi. VI. 2.
Isatis, VI. xxix.
Juglans, IV. xiv. xviii. V. 15.
Juniperus, VI. iii. IV. 28.

L.

Lamium, I. 23. V. 6. VI. 16.
Lapsana, VI. 15.
Lathyrus, VI. xvii. xxix.
Lavandula, VII. vi.
Laurus, III. 11. IV. 10. 29. V. 1.
Leontodon, VI. vii. IV. 3. 10. VIII. 21.
Leonorus, VII. vii.
Lepidium, V. iii. VII. xiii.
Leucoium, IV. xiii.
Libellula, V. xxi.
Ligusticum, VII. vii.
Ligustrum, V. xiv. X. 24.
Lilium, VI. xviii. VII. xx. VII. 10. 11.
Linum, VI. vii. xxiv. VII. 18.
Liriodendron, IV. 16. V. 4.
Lithospermum, VII. 6.
Lonicera, IV. xv. VII. xii. I. 11. 23. VI. 15. X. 26.
Lotus, VI. vii.
Lychnis, V. xxx. VI. xiv. xvii. IV. 22. V. 26. 30.
Lycopodium, IV. i.
Lycopsis, VIII. xii.
Lycopus, VIII. xii.
Lysimachia, VI. xxii. VII. iv. V. 25. IX. 2.
Lythrum, VII. ii. VII. 18.

M.

Malva, VII. xii. IV. 13. VI. 6. 15. VIII. 15. X. 26.
Marrubium, VIII. 7.
Matricaria, VI. xi. IV. 10. VI. 25. X. 26.
Medicago, VI. xxvii.

Meleagris, IV. xv.
Melica, VI. vii.
Melissa, VII. xxvi.
Mentha, VII. xv. VII. 22. 28. VIII. 7. 9.
Menyanthes, VI. xiii. IV. 22.
Mercurialis, V. 1. IV. 12.
Mespilus, IV. 22. VI. 5.
Mirabilis, VII. xvi.
Morus, IX. i. IV. 12. 18. V. 14. 20.
Motacilla, IV. xiii. V. iii. xv. II. 12. IV. 9. 16. 28. VIII. 26.
Myosotis, V. 13.
Myrica, V. xiv. VIII. 26.

N.

Narcissus, V. xv. IV. 1.
Noctes, V. xxiv. VI. 20.
Nymphæa, IV. xvii. VI. xiv. xvi. VI. 2.

O.

Oenothera, VII. v.
Ononis, VII. vi. VII. 11.
Ophrys, VI. xxii. V. 5.
Origanum, VII. x.
Orchis, VI. vii. xiv. xv. VI. 12.
Ornithogalum, IV. xv.
Orobus, V. xiii.
Osmunda, VI. xxiv.
Oxalis, V. xiii. IV. 16.

P.

Pæonia, VI. ii. xvi.
Papaver, VI. xxiv. VII. vii. VI. 7. 22.
Papilio, IV. vii.
Parietaria, V. xxiii.
Paris, VI. vii.
Parnassia, VI. xvi. IX. 5.
Pastinaca, VI. 22.
Pavo, VI. xxix.
Perdix, VII. 18.
Phaseolus, VII. 10.
Phellandrium, VI. 17.
Philadelphus, V. xiv. VI. xxiv.
Phyllerea, IV. 22.
Picris, VII. xii.
Pimpinella, VI. xviii. VII. 28.
Pinguicula, V. xxx.
Pinus, VI. i. III. 29. IV. 17. 18. 21. 22. 25. 29. V. 7. VI. 16.

Pisum,

INDEX.

Pisum, IV. 29. VIII. 21.
Plantago, V. xxviii. VI. v. IV. 26. VII. 2.
Platanus, IV. 14. 18.
Polygala, V. xxv. V. 15.
Polygonum, VI. 3. 7. VI. 8. VII. 13. 20.
Populus, XII. xxiii. IV. xix. xxx. V. ix. xvi. xx. III. 21. IV. 16. 17. 21. 25. V. 14.
Potentilla, V. xvi. VI. xi. xiv. xvi. xviii. V. 26. VII. 4.
Prenanthes, VII. x.
Primula, V. i. xiv. xv. III. 29. IV. 9. V. 4.
Prunella, VI. xxii. VI. 21.
Prunus, V. ix. xv. xxv. xxviii. III. 11. 29. IV. 1. 6. 13. 16. 18. IX. 20. X. ii.
Pteris, VIII. xxviii. IX. 20.
Pyrus, V. xv. xxvi. VI. ii. IV. 4. 6. 10. 14. 18. 23. V. 13.

Q.
Quercus, V. xxi. IV. 7. 13. V. 6.

R.
Rallus, IV. x.
Rana, IV. xiii.
Ranunculus, IV. xv. V. xxvi. VI. vii. xvii. VII. vi. III. 28. IV. 28. V. 14.
Reseda, VI. 25.
Rhamnus, V. xv. xxi. VI. xv. xvi. IV. 11. V. 31. X. 1.
Rheum, VI. xvi.
Rhinanthus, V. 4.
Ribes, V. xxv. VI. vii. II. 25. III. 2. IV. 3. 6. 13.
Robinia, V. xv. IV. 12. 21. VI. 17.
Rosa, V. xv. VI. xv. VII. vii. xxiv. IV. 4. V. 11. VI. 6. X. 1.
Rosmarinus, I. 5. IV. 22. VI. 24.
Rubus, V. vii. xxvii. VI. vii. xxiv. IV. 3. 28. V. 24. VI. 5. 6. VII. 11. 22. VIII. 30.
Ruta, VIII. 1.

S.
Sagina, VII. 2.
Salix, III. xix. IV. xxi. V. vii. xiii. xvi. III. 21. IV. 1. 7. 10. 17.

Salmo, V. xxi.
Salvia, VI. 13. VIII. 14.
Sambucus, VI. xxix. VII. xvii. II. 4. IV. 25. VI. 6. X. 1.
Sanguisorba, VII. vii.
Sanicula, VI. 8.
Saponaria, VII. xxii. VIII. 1.
Satyrium, VI. xx.
Saxifraga, V. xxvi. V. 6.
Scabiosa, VI. xxix. VIII. iv. VI. 12. VII. 28.
Scolopax, X. 22.
Scrophularia, VII. x. IV. 27. VI. 13.
Scutellaria, VI. xxvi.
Secale, V. xxv. VI. xviii. VIII. iv. V. 21. VI. 21.
Sedum, VI. xx. VII. vi. xiv. xv. VIII. 1. IV. 4. VII. 13. 22. VIII. 7.
Semina, V. viii. xiii.
Senecio, VI. vii. IV. 12. VIII. 26.
Serapias, VII. ii.
Serpentes, IV. vi.
Serratula, VII. xv. xxvii.
Silene, VI. 13. 28.
Sisymbrium, IV. 14. V. 30. VI. 6. VIII. 1.
Sitta, VIII. 8.
Solanum, VI. xxiv. VI. 16. VII. 11. 30. X. 1.
Solidago, V. xxiv.
Sonchus, VII. xi. xxi. VII. 28.
Sorbus, VI. iii. IV. 1. 13. V. 6. VI. 3. IX. 20.
Spartium, IV. 18. 22. 23. VI. 9.
Spergula, I. 26. V. 31.
Spinacia, VII. 8.
Spiræa, VII. v. IV. 22. VI. 12. 27. VII. 11.
Stachys, VI. xxv. VII. ix. VI. 9. 22.
Stellaria, IV. 22. V. 20. VI. 6.
Sturnus, V. vii.
Syringa, V. xiv. VI. xviii. xxv. IV. 9. 27.

T.
Tamus, V. 20. 30. X. 1.
Tanacetum, VII. xv. III. 29. VII. 8. VIII. 5.

Tegments,

INDEX.

Tegmenta, IV. xi.
Tetrao, II. 22.
Teucrium, VII. x. VII. 8.
Thymus, VI. 3.
Tilia, V. xxi. VII. xi. IV. 12.
   18. V. 15. VII. 13.
Tinunculus, IV. xiii.
Thalictrum, VII. 2.
Tipula, IV. vi.
Tordylium, VI. xxvii. VI. 21.
   VII. 6.
Tormentilla, VI. ii. IV. 22. V. 5.
Trifolium, VII. x. V. 27. VIII. 1.
Tringa, IV. vi.
Triticum, IV. 3. VI. 12. 21. 25.
   VII. 2. 13. VIII. 21.
Trollius, V. xxvi.
Tulipa, V. xxv.
Turdus, II. 16. III. 4. IV. 23.
   IX. 25. X. 7.
Tussilago, IV. xii. xxx.

V.

Vaccinium, V. xxx. VI. vii. xxix.
Valantia, V. 1.
Valeriana, VI. xvii. V. 4. VI. 12.
Vaporaria, IV. xix.
Verbascum, VII. i. ii. VI. 22. VII.
   11.
Verbena, VI. 17.
Veronica, V. xxx. VII. v. III. 26.
   IV. 26. V. 30.
Viburnum, V. xiv. VI. xiv. I. 23.
   III. 11. 29. IV. 4. 11. 16. VI.
   2. VII. 22. X. 1.
Vicia, VI. xxii. xxiv.
Viola, V. iii. III. 28. IV. 18. V.
   20.
Vipera, VIII. 7.
Vitis, IV. 15. VI. 6. 25.
Ulex, IV. 24. VII. 6. IX. 16.
Ulmus, V. viii. xv. IV. 1. 10. 12.
Urtica, IV. 25. VIII. 7.

INDEX.

A.

Abele, IV. xix. IV. 16. 17.
Acacia, V. xv. IV. 12. 21. VI. 17.
Adonis, VI. xiii. IX. i.
Agrimony, VII. vi. xv. VI. 24. VI. 17. VII. 18.
Alder, V. xiv. xxi. VI. xvi. III. 26. IV. 7. 11. V. 31. IX. 16. X. 1.
Ameranth, VII. 24.
Anemone, V. iii. IV. 10. V. 20.
Apple tree, V. xv. VI. ii. IV. 4. 14. 23.
Apricot, IV. 1. 6. 16. 22.
Archangel, I. 2. 3. VI. 16. 25.
Arsmart, VI. 3. VII. 20.
Ash, V. xxi. IV. 18. 22. V. 6. 20. IX. 20. 22.
Asp, XII. xxiii. IV. xix. V. xx. III. 21. IV. 25. V. 14. X. 1. 22.
Asparagus, VII. 16.
Asphodel, VII. ii.
Assarabacca, VI. ii. xiii.
Avens, VI. ii. xiii. V. 28.

B.

Barberry, V. xiii. VI. viii. IV. 12. X. 1.
Barley, IV. xvi. V. xiii. VII. vii. VIII. xvi. V. 6. VI. 28. VIII. 3.
Basil, VII. x.
Bay, III. 11. V. 1. IX. 10. 29.
Beam tree, VI. xv.
Bean, IV. 29. VII. 20.
Bees, III. 11. IV. 10.
Bedstraw, VII. vi. VII. 8. VIII. 1.
Beech, V. xvi. IV. 21. VII. 2.
Bellflower, VI. xi. VII. 11.
Betony, VII. x. IV. 27. V. 5. VI. 13. VII. 4.
Bilberry, VI. vii. xxix. VII. viii.
Bindweed, VII. i. iv. IV. 25. VI. 25. VII. 27.
Birch, V. xiii. IV. 1. 27. IX. 11.
Birds, VIII. iv.
Bittern, IV. 4.

Black cap, IV. 28.
Blackbird, IX. 27. X. 7.
Bluebottle, VI. xxii. VI. 28.
Bramble, V. vii. VI. vii. IV. 3. V. 24. VI. 5. VII. 22. IX. 11. 30.
Brass nights, VI. xxi.
Bream, V. xxx.
Briar, VI. xv. VII. vii. IV. 4. X. 1.
Brooklime, V. 30.
Broom, IV. 18. 22. 23. VI. 9.
Bryony, VI. xxviii. IV. 27. V. 20. 30. X. 1.
Buckbean, IV. 22.
Buckwheat, VI. 7.
Bugle, V. xxx. IV. 29.
Bugloss, V. 12. VI. 6.
Bur, IV. xxx.
Burdock, VII. ix. VII. 28. VIII. 12.
Burnet, VII. vii.
Buttercups, VI. 14.
Butterfly, IV. vii.
Butterwort, V. xxx.

C.

Camomile, VI. xi.
Campion, VI. xiv. xxviii. IV. 22. V. 27.
Carrot, VII. vi. VII. ii.
Catchfly, V. xxx. VI. xiii.
Cats foot, V. ix. xxvii.
Cedar, IV. 17.
Celandine, V. xxvii. V. 5. 10.
Chaffinch, II. 16. IX. 16.
Cherry tree, V. xv. xxv. xxviii. III. 29. IV. 8.
Chervil, VII. vii. V. 28. VI. 5.
Chesnut, V. xvi. IV. 1. 16. 18. V. 9. 12. X. 6.
Chickweed, VIII. i. I. 26.
Christopher herb, VI. vii.
Cicely, V. xxv. IV. 25.
Cinquefoil, V. xvi. VI. xi. xiv. xviii. xx. VI. 12.
Cistus, VI. xx.
Clary, VIII. 14.
Cleavers, IV. 3. V. 13.

Clover,

INDEX.

Clover, VII. x. V. 27. VI. 27.
Cockle, VII. v. VI. 13.
Cold, XII. v. xv.
Coltsfoot, IV. 12.
Columbine, VI. vii. V. xxv.
Comfrey, V. vii.
Cowslips, V. 4.
Crab, IV. 23.
Crakeberry, IV. xxx.
Cranes bill, V. xxvii. xxx. VI. v. ix. xi. IV. 27. V. 15.
Cresses, V. 30.
Crow, X. 2.
Crowfoot, VI. vii. IV. 28. V. 14.
Cuckow, V. xii. VII. xv. IV. 17.
Cuckow flower, V. xxviii. IV. 18. V. 20. 30.
Curlew, VIII. 14.
Currants, V. xxv. VII. ix. II. 25. III. 29. IV. 3. 13. 25. VI. 30.

D.
Daffodil, V. xv. IV. 1.
Daisy, VI. xvii. I. 26.
Dakerhen, IV. x.
Dandelion, IV. x. IV. 3. 10.
Darnel, VI. vii.
Devils bit, VIII. iv. VII. 28. VIII. 21.
Dewberry, VI. xxiv.
Dove, III. 5. VII. 30. X. 10.
Dropwort, VI. xxii.
Duck, VI. vii.

E.
Earth nut, V. 20.
Elder, V. iv. VI. xiv. xxix. VII. xvii. II. 4. III. 29. IV. 11. 12. 17. 25. VI. 2. 6. X. 1. 9.
Elecampane, VII. xxiv. VII. 24.
Elm, V. viii. xv. IV. 1. 10. 12.

F.
Fennel, VII. 9.
Fern, VIII. xxviii. IX. 20.
Feverfew, VI. xxix. IV. 10. V. 25.
Fieldfare, IV. 23. IX. 25.
Figtree, IX. iv. IV. 6. 21. 22. VII. 11.
Filberd, V. ix. IV. vii.
Fir, VI. i. IV. 18. 21. 22. 25. V. 4. 7. VI. 16. IX. 11.
Flax, VI. vii. xxiv. VII. 20.

Fleabane, VII. iv. xiii.
Flixweed, IV. 14. VI. 6.
Flower de luce, VI. 2.
Foxglove, VII. iv.
Frog, IV. xi. xii.
Furze, IV. 22. 24. VII. 6. IX. 16.

G.
Game black, IV. vi.
Gentian, V. xxii.
Germander, V. xxx. IV. 26.
Gladdon, VII. vi.
Goat sucker, VI. 5. VIII. 15. 17.
Golden rod, VI. xxiv. VII. xvi. VIII. 26.
Goldilocks, V. xxvi.
Gooseberry, II. vii. xx. 25. IV. 6. 13. 25.
Goule, V. xiv. VIII. 26.
Grass, IV. xv. V. xx. VI. iv. vii. xvi. xviii. xxviii. VII. v. xxii. III. 21. V. 13. VI. 7. 21. VII. 2. 13. IX. 5.
Greenhouse, V. viii. xxiv. VIII. xxviii. X. vi.
Gromil, VII. 2.
Groundsel, V. xvii.

H.
Harebells, IV. 26.
Harts tongue, VII. xxii.
Hasel, IV. xii. I. 23. II. 22. IX. 10.
Hawkweed, VI. xxvii. xxix. VII. xvi.
Hay harvest, VII. vii. xvi.
Heat, VI. xxix.
Heath, VI. ix.
Hedges, VI. xxi.
Hellebore, VII. ii.
Hemlock, VI. 16. 17.
Hemp, VII. xi.
Hen, VII. xx.
Henbane, VI. xii. V. 26.
Herring, IX. 17.
Holly, I. 23. IV. 3. 24. 26. X. 1.
Honeysuckle, IV. xv. V. xxx. II. 23. X. 26.
Hops, VII. vi. IV. 14. IX. 20.
Horehound, VII. ix. x. VI. 22. VIII. 7. 12.

Hornbeam,

INDEX.

Hornbeam, V. vi. III. 29. IV. 10. 13. 25.
Hot beds, IV. xi. xix.
Hounds tongue, IV. 10.
Hyssop, VII. xi.

I.
Jack by the hedge, V. xxv. IV. 10. 17.
Ice, V. 4.
Jessamine, IV. 28. VII. 11.
Iron nights, VIII. xxviii.
Juniper, IV. 8. 28.
Iva, X. xxviii.
Ivy, V. 26. VII. 9. IX. 11.
Ivy, ground, IV. 16.

K.
Kestrell, IV. xii.
Knapweed, VI. xxii. VII. i. vi. xxii. VI. 28. VII. 13.

L.
Laburnum, IV. 13.
Lady's mantle, V. xxv.
Langue de boeuf, VII. xvi.
Lapwing, IV. vi.
Larch, III. 29.
Lark, III. xxix. II. 4. IX. 25. X. 10. 14. 24.
Larkspur, VI. 23.
Lavender, VII. 6.
Laurel, III. 11. IV. 13. 16. IX. 20.
Laurustinus, I. 23. III. 1. IV. 4. VII. 22.
Lead nights, V. xxiv.
Lilac, V. xiv. VI. viii. xxv. IV. 7. 9. 27. VI. 2.
Lilly, IV. xvii. V. xxx. VI. xiv. xvi. xviii. VII. vii. V. 3. VII. 10. 11.
Lilly day, VI. xx.
Lime, V. xxi. VII. ii. IV. 12. 18. VII. 13.
Liquorice, VI. xxiv.
Loosestrife, VI. xxii.
Liverwort, VI. xvi.

M.
Mallow, VII. xii. IV. 13. VI. 6. 15. VIII. 15.
Maple, IV. 6. 18. V. 11.
Marjoram, VII. x. xi.

Marvel, VII. xvi.
Marygold, V. xx. VI. 29. IV. 16. VI. 23.
Mayweed, VI. xxiii. VI. 6.
Meadow sweet, VII. v. IV. 22. VI. 12. 27.
Medic, VI. xxvii.
Melilot, VIII. i.
Mercury, V. i. IV. 12. 24.
Mezereon, IV. xv.
Milkwort, V. xxv. V. 15.
Milleria, XI. ii.
Mint, VII. xv. VII. xxviii. VIII. vii. ix.
Moneywort, VI. iv. V. 25.
Moonwort, vi. xxiv.
Moscatel, V. viii.
Mosses, IV. i. XI. 5.
Motherwort, VII. vii.
Mugwort, V. 1. VII. 16.
Mulberry, IX. i. IV. 18. V. 14. 20.
Mullein, VII. i. ii. VI. 22. VII. 11.
Mustard, V. 16.

N.
Narcissus, IV. 1.
Nettle, VI. xxv. IV. 25. V. 6. VI. 9. VIII. 7.
Nightingale, V. xv. IV. 9. VI. 15.
Nightshade, VI. xviii. xxiv. VII. x. V. 18. 21. VII. 4. 30. X. 1.
Nuthatch, VIII. 7.

O.
Oak, V. xxi. IV. 7. 18. V. 6. 20.
Oats, VI. 28.
Onion, VIII. v.
Orach, IX. xxii. VIII. 17.
Orange, V. xiv. VI. xxiv. VII. viii.
Orchis, VI. vii. xv. xx. xxii.
Orpine, VIII. i. VIII. 7.
Osier, V. vii. xiii.
Oxeye, VI. xxix.

P.
Paris herb, VI. vii.
Parsley, VI. 21. VII. 6. 10.
Parsnep, VII. i. III. 28. VI. 3. 13. 22. VIII. 1.
Partridge, II. 22. VII. 18.

Peach,

INDEX

Peach, IV. 6. 16.
Peacock, VI. xxix.
Pear, V. xxvi. IV. 6. 18.
Pearlwort, VII. 2.
Peafe, V. xxviii. VI. xxix. IV. 29. VIII. 21.
Peiony, VI. i. xvi.
Pellitory, V. 23.
Pennyroyal, VII. 22.
Pennywort, VI. 27.
Pepper, wall, VI. xx. VII. xv.
Phyllerea, IV. 22.
Pike, IV. x.
Pilewort, IV. xv. III. 28.
Pimpernel, VI. 2.
Pine, VII. x.
Pink, VI. xvii. xxix. VII. viii. VII. 7.
Plantain, V. xxviii. VI. v. IV. 26. VII. 1. 24.
Plane tree, IV. 14. 21.
Plumb tree, V. xxviii. IV. 6. 16.
Polyanthus, V. i.
Pondweed, VI. viii.
Poplar, IV. xxx. V. xvi. IV. 21. 22.
Poppy, VI. xxiv. VII. vii. VI. 7. 22.
Potatoe, VII. xi.
Primrose, V. xv. VII. v. III. 29.
Privet, V. xiv. X. 24.

Q
Quicken tree, VI. iii. IV. i. xiii. V. vi. VI. iii.
Quince, IV. 10. 28. V. 13.

R
Radish, VI. 3.
Ragweed, IV. 12.
Rampions, VI. xvi. VII. i.
Ramsons, V. 6.
Rasberry, IV. 13. 28. VI. 6. 24. VII. 11.
Rattle VII. vii. V. 4.
Redstart, IV. 16.
Rest harrow, VII. vi. VII. 11.
Rhapontic, VI. xvi.
Robert herb, VI. xiv. IV. 23.
Robin, VIII. 26.
Roche, V. xxv.
Rocket, VI. 25.

Rook, II. 12. III. 2. VI. 21. VIII. 12. 17. X. 7.
Rose, V. xv. VI. xv. VII. vii. VIII. vii. V. 11. VI. 6.
Rosemary, I. 5. IV. 22. VI. 24.
Rue, VII. iii. VIII. 1.
Rupture wort, V. xxix.
Rush, V. xxx.
Rye, V. xxv. VI. xviii. VIII. iv. V. 21. VI. 21.

S
Saffron, IV. xii. VII. xviii. VIII. xxviii.
Sage, VII. viii.
Saint John's wort, VI. xxix. VII. vi. xii. IV. 27. VI. 22.
Sallow, III. xix. V. xvi. X. xxviii. III. xi. IV. 6. 7. X. 5.
Salmon, V. xxi. xxviii.
Sampire, IX. xvii.
Sanicle, VI. 8.
Saw-wort, VII. xv.
Saxifrage, V. iii. xxvi. VI. xviii. V. 6. VI. 9. VII. 28.
Scabius, VI. xxix. VI. 12.
Seeds, IX. xiv.
Self heal, VI. xxii. VI. 21.
Serpents, IV. vi.
Sheep, V. iii.
Shoots, VI. 28. VII. 11.
Silverweed, VII. 4.
Smallage, VIII. 29.
Smelt, IV. xxi.
Sneezewort, VI. xxix. VII. vii.
Snow, V. xviii. IV. 29.
Snowdrops, IV. xii. I. 26.
Soapwort, VII. xxii. VIII. 1.
Solomon's seal, V. 10.
Sorrel, V. xiii. IV. 16.
Sowing, V. viii. xiii.
Spearwort, VI. xvii. VII. 6.
Speedwell, VII. v. III. 26.
Spider, IV. vi.
Spikenard, VII. xxiii.
Spindle tree, V. xiv. X. 9.
Spurge, IV. xxv.
Spurrey, I. 26. V. 31.
Star of Bethlehem, IV. xv.
Starling, V. vii.
Steel nights, II. xxii.

Stitchwort,

INDEX.

Stitchwort, IV. 22. V. 20. VII. 6.
Stonecrop, VII. vi. xiv. VII. 13.
Strawberry, VI. xxvi. IV. 13. VII. 9.
Succory, VII. xii. VI. 28.
Sundew, VII. x.
Swallow, V. ix. IX. xvii. IV. 6. IX. 21.
Swan, V. ix. IX. xvii.
Sycomore, IV. 13. 25.

T.

Tansey, VI. xvi. VII. xv. III. 29. V. 26. VII. 8. VIII. 5.
Tare, VI. 6.
Teasel, VII. xviii. VII. 29.
Thermometer, II. 20. 25. 26. III. 2. 28. IV. 20. 29. V. 4. 30. VI. 19. 25. VII. 8. 15. VIII. 8. 16. IX. 10. 16. X. 6.
Thistle, VI. xxii. VII. ix. x. xi. xvii. xxi. xxvii. VII. 29. VIII. 12.
Thorn, V. iv. xv. xxviii. VI. xv. xvii. III. 29. IV. 18. 22. 28. V. 10. VI. 5. X. 2.
Throatwort, VI. xxix. VII. ii. vi.
Thyme, VI. 3.
Thrush, II. 16. III. 4. IX. 29.
Toad flax, VI. xxviii.
Tormentil, VI. ii. IV. 22. V. 5.
Touch me not, VII. xviii.
Travellers joy, IX. 2.
Trefoil, VI. vii.
Tulip, V. xxv.
Tulip tree, IV. 16. V. 4.
Turkey, IV. xv. V. 15.

Turneps, IV. 15.
Tway blade, VI. xxvi. V. 5.

V.

Valerian, VI. xvii. V. 4. VI. 12.
Vervain, VI. 17.
Vetch, VI. xii. xvii. xxii. xxiv.
Vine, IV. 15. VI. 6. 25.
Violet, IV. xii. V. iii. vii. VII. xxvi. III. 28. IV. 18. V. 3. 20. VII. 2.
Viper, VIII. xxix.

W.

Wagtail, IV. xii. V. iii. IX. xvii. II. 12.
Wallflower, IV. 2.
Wallnut, IV. 14. V. 15.
Wheat, IV. 3. VI. 12. 21. 25. VII. 11. VIII. 21.
Wheat ear, V. iii.
Whorts, V. iii.
William, sweet, V. xxiv.
Willow, IV. xxi. IV. 1. 7. 10. 17. VII. 11.
Willow herb, VI. xxii. xxiv. xxvi. VII. i. ii. xii. VI. 13. VII. 18. 28. VIII. 21. IX. 2.
Woad, VI. xxix. VI. 25.
Woodbind, VII. 12. VI. 15.
Woodcock, X. 22.
Wormwood, VIII. xiv. IX. vi. IV. 9. VIII. 9.

Y.

Yarrow, VI. xxiii. VII. vi. VI. 18.
Yew tree, III. xxix. IV. 24.

THE
SWEDISH PAN.

THE SWEDISH PAN.

BY

NICOLAS HASSELGREN.

UPSAL, 1749. Decem. 9.

Amænit. Academ. vol. 2.

§. 1.

THE antients attributed the pastoral life to *Pan*, the care of flowers to *Flora*, hunting to *Diana*, and the cultivation of grain to *Ceres*. We, tho' acknowledging only one Deity, who governs all things, yet often use these names to denote the subject we undertake to treat upon. What word is now more known among botanists than the word *Flora*; by which they mean all those plants, which grow within a certain compass of ground; as our Fauna Suecica takes

takes in all those animals, which are natives of Sweden? For a like reason we have entitled this small tract the *Swedish Pan*; intending thereby to denote the five domestic quadrupeds, which live upon plants growing in Sweden; or the devouring army of *Pan*, which lays waste the provinces of the *Swedish Flora*. We choose by this means to avoid a prolix definition, which is always disagreeable for the title of a book.

§. 2.

The pastoral life, by the testimony of both sacred, and prophane history, is nearly as old as man himself; so that i would willingly derive the knowledge which i am going to deliver, from the most ancient times. But altho' plants have been constantly obvious to the eyes of every man; yet i am obliged to declare, that we have nothing delivered down to us in any book concerning the kinds of plants proper for the different kinds of cattle; so that i may be sure of not disgusting my reader with stale matter new dressed up. For the whole of what i present to him is new.—Our illustrious president in his journey thro' Dalecarlia ann. 1734, made the first attempt this way, as may be seen Flor.

Lapp. p. 158. where he says thus. 'In my
'journey thro' Dalecarlia, when we had climbed
'up the mountains, and were got into Norway,
'my fellow travellers being tired, and asleep,
'i wandered about in a dismal wood, and per-
'ceived that the horses easily distinguished
'wholesome from noxious food; for being very
'hungry, they devoured all sorts of plants, ex-
'cept the following; *meadow sweet, valerian,*
'*lilly of the valley, angelica, loose-strife, marsh-*
'*cinquefoil, cranes bill, hellebore, monks-hood,* and
'many shrubs. This gave me a hint to re-
'commend to the curious, that they would set
'about examining what plants such animals, as
'live on vegetables, viz. *the cow,* the *sheep,* the
'*goat,* the *deer,* the *horse,* the *hog,* the *monkey,*
'and their species will not touch. An examina-
'tion which would not be without its use, were
'it properly made.' Notwithstanding this re-
commendation, no enquiry was made, till our
president returned home from his travels thro'
forreign countries, and made a progress thro'
our own provinces. Afterwards professor Kalm,
that worthy disciple of so great a master, follow-
ed his example; so that in his journey to Bahus
we find mention made of some plants, which
cattle either eat, or refuse. Ann. 1747 and
1748

1748 our prefident undertook with great diligence not only to make experiments himfelf, but to excite his difciples, and auditors to do the fame; of which number i was one. Thus at laft many experiments were made, and repeated, efpecially by D. D. Hagftrom, Mag. E. G. Liidbeck, E. Ekelund, J. G. Wahlbom, L. Montin, F. Oldbers, J. C. Forfkahl, A. Fornander; not to mention others, who ftrove, as it were, to out-do one another in finding the plants, which were fuitable to different animals.

§. 3.

The difficulty however of examining all the Swedifh plants, and getting animals proper for experiments, which ought all to be repeated, has hindered us from being able to give a compleat work on this fubject. But the greateft part, and the moft common vegetables of Sweden being now determined by us; what is wanting may be fupplied from time to time. We hinted that animals proper for experiments, which ought to be taken from among *cows, goats, fheep, horfes*, and *fwine*, are difficult to be found, for thefe reafons; firft, becaufe fome plants are eaten by them in the fpring, which they

THE SWEDISH PAN.

they will not touch all the summer; when they are apt to grow rank in taste, and smell, and become stalky and hard. Thus many people eat the *nettle* in the spring; but who could bear it afterwards? Again, because some kinds of animals eat the flower, and will not eat the stalks; others eat the leaves and will not eat the stalks. N. B. When they eat the leaves, we say in general they eat the plant, otherwise there would be few grasses they could be said to eat. Œcon. Nat. Next, the animals ought not to be over hungry, when we make our experiments, if we intend to make them properly. For they will greedily devour most kinds of plants at such a time, which they will absolutely refuse at another. Thus when they come immediately out of the house, they are not fit to make experiments upon; for then they are ravenous after every green thing that comes in their way. The best method is to make the experiments when their bellies are almost full, for they are hardly ever so intirely. Moreover the plants ought not to be handled by sweaty hands; some animals will refuse the most pleasing and tasteful in that case. We ought to throw them on the ground, and if we find the animal refuses to eat them, we must mix them

with

with others that we know they like; and if they still refuse them, we have a sure proof; especially if the same be tried with many individuals.

§. 4.

Our views do not extend beyond the Swedish plants, and that for the sake of our own œconomy. Let forreigners look to that part which concerns themselves, and thus our work will be confined within moderate bounds. We can produce above 2000 certain experiments, some of which were repeated ten times over, some twice as often. If we take the *Flora Suecia* Holm. 1745, and put to any herb the generical name, adding the number, and some epithet by way of difference, our work will be very much abridged.

§. 5.

It is manifest that the vegetable world was intended for the support of the animal world; insomuch that altho' not a few animals are carnivorous, yet these animals which they devour cannot subsist without vegetables. In this speculation

culation we behold with admiration the wisdom of the Creator, which has made some vegetables absolutely disagreeable to some animals that live upon plants, while these plants are agreeable to others. And there are plants, which are poisonous to some animals, which are very wholesome to others, and on the contrary. This did not happen by chance, but was contrived for wise purposes. For if the Author of nature had made all plants equally grateful to all kinds of quadrupeds, it must necessarily have happened, that one species of them being remarkably increased, another species must have perished with hunger, before it could have got into better pasture; the vegetables being consumed over a large tract of ground. But as it is ordained every species must by force leave certain plants to certain animals, so that they always find something to live upon, till they meet with better pasture; in the like manner we find it contrived in relation to the plants themselves, which do not all grow in the same countrey, and climate; but every plant has its place appointed by the Creator, in which it grows more abundantly, than any where else. From hence we may observe, that those animals, which chiefly live upon particular plants, chiefly

abound

abound in certain places. Thus the *lichen* or *liverwort*, Fl. 980. is found in greatest plenty on the cold alps, and therefore the *rhen deer*, which all winter live mostly upon this plant, are obliged to live there. The festuca, Fl. 94. which florishes and spreads most on dry pastures, draws the sheep thither, which above all things delight in that kind of grass. The seeds of the *dwarf birch*, Fl. 777. which afford the best sort of food to the *rough-legg'd partridge*, and the *Norway rat*, Fn. 26. tempt them to dwell in these northern parts of the world. *Camels hay*, Mat. Med. 312. which above all plants, thrives on loose sand, draws the camel to choose those barren places, as they there find food most agreeable to them; not to mention many other similar instances. Trees, whose heads shoot up so high, that quadrupeds cannot easily reach them, afford nourishment for that reason to more numerous tribes of insects, as the *fallow*, the *oak*, the *pear*, &c. The Creator, who most wisely established this law, has as it were imprinted it on the organs of animals, that they might not offend against it thro' ignorance; and as every transgression has its punishment allotted, so also no offence against the law of nature

nature can escape. Animals, which violate this law, are punished by diseases or death; and hence we behold with admiration that brutes, which were designed to be guided by instinct, can by no means whatever be prevailed upon to act against it. If by chance it happens that any animal offends this way, and suffers for it, we vulgarly say it has taken poison; so that ignorant people wonder, not to say murmur at the wise disposition of the Creator, who has produced so many noxious plants; but without sufficient reason, for no one plant in the world is universally poisonous, but all things are good, as they came from the hands of the Creator. Physicians often mention that this or that plant is deadly, because its particles are of a nature apt to wound the fibres of the body or corrupt the juices. But this is only respectively to the species of animals, e. g. the *sun-spurge*, Fl. 536. has a milky juice, which causes blotches in our skin and hurts our fibres, and therefore it is said to be poisonous; yet the *moth*, Fn. 825. almost entirely lives upon this plant, and prefers it both for taste and nourishment to all others, as it thrives best upon it. Thus one animal leaves that, which to itself is poisonous, to another animal, which feeds upon it deliciously.

Long-

Long-leaved water hemlock will kill a *cow*, whereas the *goat* browses upon it greedily. *Monks-hood* kills a *goat*, but will not hurt a *horse*; and the *bitter almond* kills a *dog*, but is wholesome food for *man*. *Parsley* is deadly to *small birds*, while *swine* eat it safely; and *pepper* is mortal to *swine*, and wholesome to *poultry*. *Thus every creature has its allotted portion.* Animals distinguish the noxious from the salutary by smell and taste. Younger animals have these senses more acute, and therefore are more nice in distinguishing plants. An empty stomach will often drive animals to feed upon plants, that were not intended for them by nature. But whenever this has happened they become more cautious for the future, and acquire a certain kind of experience; e. g. *the monks-hood*, which grows near Fahluna, is generally left untouched by all the animals, that are accustomed to these places; but if forreign cattle are brought thither and meet with this vegetable, they venture to take too large a quantity of it, and are killed [b]. The cattle that have

---

[b] The same thing has been told me by the countrey people in Herefordshire in relation to *meadow-saffron*, which grows in plenty in some parts of that county. Gmelin, Fl. Sibirica, p. 76. says that cattle eat the leaves of the *hellebore*, 40. when they first spring out of the ground, and are thereby killed.

been reared in the plains of Schonen, and Weſtrogothia, commonly fall into a dyſentery when they come into the woodland parts, becauſe they feed upon ſome plants, which cattle uſed to thoſe places have learned to avoid. In the ſpring, when the *water hemlock* is under water, ſo that the cows cannot ſmell it, they dye in heaps [e]. But when the ſummer comes on

---

[e] This affair is of ſo much conſequence to the farmer, that i think it right to tranſcribe a paſſage out of Linnæus upon this ſubject.

"When I arrived, ſays he, at Tornea, the inhabitants complained of a terrible diſeaſe, that raged among the horned cattle, which upon being let into the paſtures in the ſpring, dyed by hundreds. They deſired that I would conſider this affair, and give my advice what was to be done in order to put a ſtop to this evil. After a proper examination, i thought the following circumſtances worth obſerving.

1. That the cattle dyed as ſoon as they left off their winter fodder, and returned to grazing.
2. That the diſeaſe diminiſhed as the ſummer came on, at which time, as well as in the autumn, few dyed.
3. That this diſtemper was propagated irregularly, and not by contagion,
4. That in the ſpring the cows were driven into a meadow near the city, and that they chiefly dyed there.
5. That the ſymptoms varied much, yet agreed in this, that the cattle, upon grazing indiſcriminately on all ſorts of herbs, had their bellies ſwelled, were ſeized with convulſions, and in a few days expired with horrible bellowings.

6. That

on and has dryed the ground, they are very careful not to touch it. It is also true, that all vegetables prohibited by nature to particular animals are not equally pernicious; and therefore though through necessity and hunger they eat

6. That no man dared to flay the recent carcases, as they found by experience, that not only the hands of such as attempted it, but their faces too had been inflamed, and mortified, and that death had ensued.

7. The people enquired of me, whether there were any kinds of poisonous spiders in that meadow, or whether the water which had a yellowish tint was not noxious.

8. That it was not a murrain, was clear, because the distemper was not contagious, and because that distemper is not peculiar to the spring. I saw no spiders but what are common all over Sweden; and as to the water, the sediment at the bottom, that caused the yellowness, was nothing but what came from iron.

9. I was scarcely got out of the boat, which carried me over the river into the meadow, before i guessed the real cause of the disease. For i there beheld the *long-leaved water hemleck*. My reasons for guessing this were as follow.

10. Because in that meadow, where the cattle first fell ill, this poisonous plant grows in great plenty, chiefly near the banks of the river. In other places it was scarce.

11. The least attention will convince us that brutes shun whatever is hurtfull to them, and distinguish poisonous plants from salutary by natural instinct; so that this plant is not eat by them in the summer, and autumn, which is the reason that in those seasons few cattle dye, viz. only such as either accidentally, or pressed by extreme hunger, eat of it.

12. But

eat them, yet they do not immediately dye; but it is certain that they cannot have from thence good and proper nourishment.

§. 6.

The end of this kind of knowledge is not

12. But when they are let into the pastures in spring, partly from their greediness after fresh herbs, and partly from the emptiness and hunger which they have undergone during a long winter, they devour every green thing which comes in their way. It happens moreover that herbs at this time are small, and scarcely supply food in sufficient quantity. They are besides more juicy, are covered with water, and smell less strong, so that what is noxious, is not easily discerned from what is wholesome. I observed likewise, that the radical leaves were always bitten, the others not; which confirms what i have just said.

13. I saw this plant in an adjoining meadow mowed along with grass for winter fodder; and therefore it is not wonderfull, that some cattle, tho' but a few, should dye of it in winter.

14. After i left Tornea i saw no more of this plant till i came to the vast meadows near Limmingen, where it appeared along the road, and when i got into the town i heard the same complaints, as at Tornea, of the annual loss of cattle with the same circumstances.

15. It would therefore be worth while to eradicate carefully these plants, which might easily be done, as they grow in marshy grounds; and are not hard to find, as they grow by the sides of pools or rivers. Or if this could not be done, the cattle should not be suffered to go into such places, at least during the spring. For i am persuaded, that later in the year they can distinguish this plant by the smell alone.

bare,

bare curiosity, although were this the case every part of knowledge, which sets forth the stupendous works of the Creator, is never to be looked upon as of no consequence. On the other hand, we do not pretend to gain any medicinal advantages from these speculations, namely, to be able from hence to conclude, that this or that plant is noxious to man, because it is so to this or that brute animal. Nor do we for that reason approve of Wepfer's experiments upon dogs, and other animals, as if any knowledge can be thence gained in regard to man. No, the end we aim at is merely œconomical.

α. From these experiments we may know whether certain pastures afford good nourishment for this or that species of animals. We see e. g. *heifers* waste away in enclosures, where the *meadow-sweet* grows in abundance, and covers the ground so that they can scarce make their way through it; the countrey people are amazed, and imagine that the pasture is too rich for them; not dreaming that the *meadow-sweet* affords them no nourishment. Whereas the *goat*, which is bleating on the other side of the hedge, is not suffered to go in, though he longs to be browsing upon this plant, which to him is a most delicate and nourishing food.

β From

THE SWEDISH PAN.

β. From these experiments we may almost be sure by affinity and analogy, whether meadows or pastures are salutary or noxious to particular animals; e. g. long experience has taught us that our sheep take up poison in marshy grounds, though no one till lately knew what was the particular poison. Yet the *spiderwort* 267. the *mouse-ear scorpion grass* 149. the *mercury* 823. the *sun-dew* 257,8. the *hairy wood grass* 287. the *lesser spearwort* 458. the *butterwort* 21. have evidently suspicious marks [c]. I will therefore propose a new experiment. The *andromeda* Fl. Virgin. 160. is known to be a most rank poison to sheep in Virginia. The *andromeda*, called by the people of New York *dwarf laurel*, Cold. Act. Upsal. 1743. p. 123. is very fatal to the sheep in New York. These two plants are of a different species, but of the same natural genus, and therefore have the same vertues. Amongst us, especially in the northern parts, the wild *rosemary, andromeda*

[c] There is great reason to think that what makes low grounds so noxious to sheep is not the moisture, but the plants that grow there. For it is observed by shepherds that the great danger to sheep is immediately after a fresh spring of grass, which i imagine is owing to their licking up the young and tender shoots of poisonous plants, along with their proper food, not being able to distinguish them.

335. grows every where in marshy grounds, which being of the same natural genus with the foregoing, we may reasonably conclude that it destroys our sheep. To this we may add, that it is on account of three other species of *andromeda* 336,7,8. which grow on the Lapland mountains, that the sheep there never are healthy; and lastly although the *cistus ledon* 341. is not a species of *andromeda*, yet being of the same natural class, it is not unlikely but that this plant is far from affording good nourishment to sheep. This conjecture gives our shepherds an unexpected opportunity of making experiments with their sheep; and indeed they cannot omit to do it without being justly blameable, since on this the health of their whole flock depends. It is particularly to be noted upon this occasion, that the botany of America, a countrey so far disjoyned from us, gives a hint for considering things of the greatest use, of which the antients did not so much as dream.

7. From hence the œconomist may truly judge of his meadows, and know that some are vastly preferable to others for certain animals. For although cattle, pressed by necessity and hunger, will feed upon vegetables less gratefull to them; yet it is not to be doubted but that

they

THE SWEDISH PAN.

they are not equally well nourished by these as by others. Thus the Dalecarlians are obliged in a scarcity of wheat to support themselves by *bread* made of the *bark* of the *pine*; yet it does by no means follow from hence that this affords proper nourishment. We see that *horses* in time of war, when pressed by extreme hunger, will eat *dead hedges*, but we cannot hence conclude, that wood is good food for them.

ſ. The industrious farmer may judge from hence, when he sows his meadows with hay seeds for pasture, that it is not indifferent what kinds of seeds he chooses, as the vulgar think. For some are fit for *horses*, others for *cows*, &c. *Horses* are nicer in choosing than any of our cattle; *siliquose* and *siliculose* plants particularly are not relished by them. *Goats* feed upon a greater variety of plants than any other cattle, but then they chiefly hunt after the extremities and flowers. *Sheep* on the contrary pass by the flowers and eat the leaves. Not to mention the different disposition in different animals as to grazing near the ground or not. The countreyman who understands these things, and knows how in consequence to dispose of his grounds, and assign each kind of cattle to its properest food, must necessarily have them more healthy and fat, than he who is deftitute of thefe principles. The good œconomift will obferve the fame of his hay. For although many herbs, when dry, are eat, which when green would be refufed, it does not follow from hence that they yield good nourifhment. Much might be added concerning the propenfion of cattle to this or that plant, which the compafs of this fmall tract will not admit of; e. g. that *fheep* above all things delight in the *feftuca* 95. and grow fatter upon it than any other kind of grafs; that *goats* prefer certain plants, but being led by an inftinct peculiar to themfelves, they fearch more after variety, and do not long willingly ftick to any one kind of food whatever; that *geefe* are particularly fond of the feeds of the *feftuca*, Fl. 90; that *fwine* greedily hunt after the roots of the *bull-rufh* 40. while they are frefh, but will not touch them when dry. Hence it appears that it is in vain to contrive engines to extract the roots of the *bull-rufh* out of the water, and dry them for the ufe of thefe animals in winter. Becaufe thefe animals fpoil the meadows, where the *fcorzonera* grows, in order to come at its root, which they delight in; and alfo the fields, to get at the roots of *clowns-all-heal*, the hufbandman imagines they do do good to his fields by ploughing the ground and eating the roots of *couch-grass*, whereas they never touch them, but when pressed by the utmost necessity [d].

§. 7.

To give a view of my design in a few words. I have disposed the plants mentioned in the Flora Suecica according to their numbers; and to be as short as possible, it was necessary to add the generical name with a short and incompleat

---

[d] In the same way with us it is a notion that prevails commonly that *cows* eat the *crow-foot* that abounds in many meadows, and that this occasions the butter to be yellow, from whence i suppose it is generally known by the name of the *butter-flower*. But this i believe is all a mistake, for i never could observe that any part of that plant was touched by *cows* or any other cattle. Thus Linnæus observes, Fl. Lapp. p. 195. that it was believed by some people that the *marsh marygold* made the butter yellow, but he denies that cows ever touch that plant. Yet he thinks that all kinds of pasture will not give that yellowness, and then observes that the best and yellowest butter he knows, and which is preferred by the dealers in those parts to all other butter, was made where the *cow-wheat* grew in greater plenty than he ever saw any where else. This shews how very incurious the countrey people are in relation to things they are every day conversant with, and which it concerns them so much to know.

epithet, which however may be illuftrated out of the Flora itfelf. I have diftinguifhed the cattle againft every plant into five columns. The firft of which contains *oxen*. The fecond *goats*. The third *fheep*. The fourth *horfes*. The fifth *fwine*. By the mark (1) i have denoted thofe plants which are eaten; by the mark (0) thofe which are not eaten; by both together thofe which are fometimes eaten, fometimes refufed; or are eaten when cattle are more ufed to them, and are more hungry, otherwife not.

§. 8.

Upon the firft view of this fubject the reader will perceive, that it is not treated compleatly, fo that every Swedifh plant is pointed out, and by what animals it is eaten. What generally happens upon breaking up old pafture lands, viz. that for the firft years it cannot be cleanfed from all ufelefs weeds, and be laid down fine like a garden, but will here and there have rough tumps and hard clods, unlefs we will let it lye fallow for a very long time; the fame or fomething like it has happened upon this occafion.

I am

THE SWEDISH PAN.

I am apt to believe, however, that the reader will be better pleased that i have opened this new scene, than if i had waited longer in order to gain farther light. For since there are many people here curious in botany and œconomy, i hope they will all lend a helping hand, that i may one day be enabled to give a more compleat edition of this piece*.

---

* After this in the original follows a long table of experiments, of which i shall only give a small specimen; as the whole would increase the bulk but not the value of this piece to such readers as this translation is intended for, since they would neither know the plants by the names the author has given them, nor by any i could put in their room. However i shall for curiosity give a specimen, and add the general result of his experiments, just as he has marked it at the end of his table; which is as follows. 'Thus far,' says he, 'we have given 2314 experiments. From these it appears that

Oxen eat	276	refuse	218 plants
Goats	449		126
Sheep	387		141
Horses	262		212
Swine	72		171

'And thus these animals leave untouched 886 plants.
'These animals will not eat any kind of *moss*.
'The goats are very fond of the *algæ*.
'Some of them greedily devour the *fungi*, others will not taste them. But we recommend farther trials in relation to these matters.'

Then follows an account of some trials made by Dr. O. Hagstrom to the same purpose in relation to *rben deer*,
but

but as they no ways concern us i have omitted to mention them.

N. B. For the table i have chosen not to take such plants as occurred first in my author, but to select the grasses of our own countrey, and have given English names to them of my own invention, the reason of which will appear in the following observations.

	O.	G.	S.	H.	Sw.
Spring grass	1	1	1	1	-
Mat grass	10	1	1	1	0
Canary grass, *reed*	1	1	1	1	0
Cat's-tail, *meadow*	1	1	0	1	0
Fox-tail, *meadow*	10	1	1	1	10
———— *flote*	1	1	1	1	0
Millet grass	1	1	1	-	-
Bent grass, *silky*	-	1	0	1	-
———— *fine*	1	1	-	1	-
Hair grass, *small leaved*	1	1	1	1	-
———— *water*	1	-	1	1	-
Meadow, *creeping*	1	1	1	1	0
———— *annual*	1	1	1	1	1
———— *great*	1	1	1	1	1
———— *narrow leaved*	1	1	1	1	1
———— *common*	1	1	1	1	1
Cock's-foot grass, *rough*	0	1	1	1	0
Dog's-tail grass, *crested*	-	-	1	-	-
———— *blue*	-	1	1	1	0
Fescue grass, *flote*	0	1	1	1	10
———— *purple*	1	1	1	1	-
———— *sheep's*	1	1	11	1	-
Brome grass, *field*	1	1	1	1	-
———— *spiked*	-	1	1	1	-
Oat grass, *meadow*	1	1	1	1	-
———— *bearded*	-	10	-	-	-

Obſervations on GRASSES.

Observations on GRASSES.

AS the foregoing treatise contains some observations on grasses [f], that are quite new, and as this affair is of the utmost importance to the husbandman, i shall subjoyn some observations of my own relating to the same subject.

It is wonderfull to see how long mankind has neglected to make a proper advantage of plants of such importance, and which in almost every countrey are the chief food of cattle. The farmer for want of distinguishing, and selecting grasses for seed, fills his pastures either with weeds, or bad, or improper grasses; when by making a right choice, after some trials he might be sure of the best grass, and in the greatest abundance that his land admits of. At present if a farmer wants to lay down his land to grass,

[f] *By grasses are meant all those plants, which have a round, jointed and hollow stem, surrounded at each joint with a single leaf, long, narrow and pointed, and whose seeds are contained in chaffy husks. It appears by this definition, which is Ray's, that all the kinds of grain, as wheat, oats, barley, &c. are properly grasses, and that the broad, the white, the hop, &c. clovers are not grasses, though so frequently called by that name.*

OBSERVATIONS ON GRASSES.

what does he do? he either takes his seeds indiscriminately from his own foul hayrick, or sends to his next neighbour for a supply. By this means, besides a certain mixture of all sorts of rubbish, which must necessarily happen; if he chances to have a large proportion of good seeds, it is not unlikely, but that what he intends for dry land may come from moist, where it grew naturally, and the contrary [g]. This is such a slovenly

[g] *Since the first edition of these tracts i have had several opportunities of observing instances of this slovenly kind of husbandry, and its effects. Instead of covering the ground in one year with a good turf, i have seen it filled with weeds not natural to it, and which never would have sprung up, if they had not been brought there.*

Arguments are never wanting in support of ancient customs, and i am no stranger to the arguments, such as they are, which prejudice and indolence have made use of on this occasion.

1. Some say then, that if you manure your ground properly, good grasses will come of themselves. I own they will. But the question is how long it will be before that happens, and why be at the expence of sowing what you must afterwards try to kill by manuring? which must be the case, as long as people sow all kinds of rubbish under the name of hay seeds. Again, if the best way is to let the ground take its chance, why is the farmer at the expence of procuring the seeds of the white, and broad clover, which come up in almost all parts of England spontaneously? but if this is allowed not to be the best way in relation to clover of any kind, what reason can be in nature, why grass seeds only ought not to be sown pure?

2. *Others*

slovenly method of proceeding, as one would think could not possibly prevail universally; yet this is the case as to all grasses except the darnel grass, and what is known in some few counties by the name of the Suffolk grass; and this latter instance is owing, i believe, more to the soil than any care of the husbandman. Now would the farmer be at the pains of separating once in his life half a pint, or a pint of the different kinds of grass seeds [h], and take care

2. *Others say, that it is better to have a mixture of different seeds. I will suppose this to be true. But cannot a mixture be had though the seeds be gathered, and separated? and is not a mixture by choice more likely to be proper, than one by chance? especially after a sufficient experience has been had of the particular virtues of each sort, the different kinds of cattle each grass is most adapted to, the different grounds where they will thrive best, &c. all which circumstances are now in general wholly unknown, though of the utmost consequence.*

3. *It is said by some, that weeds will come up along with the grass. No doubt of it. Can any one imagine that grass seeds should be exempted above from what happens to every other kind of seed. But i will venture to say, that not near the quantity of weeds will spring up which they imagine, if it be sown very thick. Men must be very much put to it, when they make such objections as this last, or indeed any of the others. I am almost inclined to say with a great writer, ' It is a simple ' thing to take much pains to answer simple objections.'*

[h] *I have had frequent experience how easy it is to gather the seeds of grasses, having employed children of ten or eleven years old*

care to sow them separately; in a very little time he would have wherewithal to stock his farm properly, according to the nature of each soil, and might at the same time spread these seeds, separately over the nation by supplying

old several times, who have gathered many sorts for me without making any mistakes, after i had once shewn them the sorts i wanted.

I have procured thus the creeping bent, the fine bent, the sheep's fescue, the crested dog-tail, &c. in sufficient quantities to begin a stock, but for want of a proper opportunity of cultivating them myself, or meeting with any one who had zeal enough to bestow a proper care on them, my collections of this kind hitherto have only proved that the scheme is in itself feasible.

This very year 1761, a little boy by my direction gathered as much of the crested dog-tail in 3 hours by the side of a road, as when shed, yielded upon weighing above a quarter of a pound averdupois, perfectly free from husks. As this seed is small the skilful will easily judge how far such a quantity would go if properly employed.

My very estimable and ingenious friend Mr. Aldworth, who was witness of the fact which i last mentioned, at my desire ordered a small part of a meadow, near his seat at Stanlake, which had better grasses and less mixed than the rest, to be left unmowed till the seeds were fit for gathering. This piece yielded upon threshing and sifting a full bushel by measure of almost pure seed of the crested dog-tail. In case any one should be inclined to follow this example, i think it highly necessary to observe that care must be taken to mow the grass before it sheds; that it be mowed very early in the morning before the dew is off the ground, and that it ought not be spread as in making hay, but left as it falls from the scythe a sufficient time, and then gently turned over.

the

the seed-shops. The number of grasses fit for the farmer is, i believe, small; perhaps half a dozen, or half a score are all he need to cultivate; and how small the trouble would be of such a task, and how great the benefit, must be obvious to every one at first sight. Would not any one be looked on as wild who should sow *wheat, barley, oats, rye, pease, beans, vetches, buck-wheat, turneps,* and weeds of all sorts together? yet how is it much less absurd to do what is equivalent in relation to grasses? does it not import the farmer to have good hay and grass in plenty? and will cattle thrive equally on all sorts of food? we know the contrary. Horses will scarcely eat hay, that will do well enough for oxen and cows. Sheep are particularly fond of one sort of grass, and fatten upon it faster, than on any other in Sweden, if we may give credit to Linnæus. And may they not do the same in England? How shall we know till we have tryed? Nor can we say that what is valuable in Sweden may be inferior to many other grasses in England; since it appears by the Flora Succica that they have all the good ones that we have. But however this may be i should rather choose to make experiments, than conjectures.

OBSERVATIONS ON GRASSES.

I now propose to add a few observations on some of our grasses, as far as i have been able to make any with some appearance of probability; but as there has reigned hitherto the greatest confusion in the English names of these most valuable plants, and as they have never been properly ranged but by Linnæus, i shall first, in imitation of that great author in his Flora Suecica, give new generical names with trivial ones to distinguish the species of all our English grasses [l]. I mean all those which are found in that author; as for the rest, since some are omitted by him, their names may be easily supplyed when their genera are settled by the learned [k]. It happens very luckily, that our common people know scarce any of the grasses by names, as far as i could ever find by conversing with farmers, husbandmen, &c. so that something may be done to remove this confusion, if a list of names be settled and agreed

---

[l] Mr. Hudson having thought proper to adopt my names with some alterations; and having cleared up many of the species of grasses in a better manner than has been done before; i have referred throughout to his Flora Britannica, which is likely to be in the hands of all who are curious in botany.

[k] This has since been done in some measure in the afore mentioned Flora Britannica.

OBSERVATIONS ON GRASSES.

on by such as are likely to have influence sufficient in these matters. As to my own list, it is only meant as a hint for others to work upon.

In giving names i have had two things in view. First to retain as much as possible such as have hitherto been used for some species of the genus. Secondly, where that could not be done, to give such as are of easy and familiar pronunciation to our common people, and at the same time approach as near as possible to the Latin names in sound where they could not be interpreted. This was done for the sake of the learned, for the more easy recollecting the botanical name. Thus i have called the *aira hair*-grass, the *bromus brome*-grass, &c. in others i have merely translated the Latin name, as alopecurus *fox-tail* grass, cynosurus *dog-tail* grass, &c.

After these preliminary observations i hope it will not be necessary to make any apology for the liberty i have taken. I am certain that till names properly adapted to the purpose be invented, we have little chance of seeing any general reformation made in this part of husbandry; and even after this without some person properly qualified to direct the countrey people, and shew them the grasses with their names,

nothing

nothing will come of that most useful doctrine delivered in the foregoing treatise of Haſſelgren [1]. But it is to be hoped that gentlemen at least will not be ſo incurious as to remain ignorant of what imports them ſo much to know. Nor is the mere botaniſt leſs concerned in the ſucceſs of this ſcheme, for there is great reaſon to think that many of the graſſes are not thoroughly ſettled, varieties perhaps being put for different ſpecies [m]; now this uncertainty can never be better cleared up than by ſowing the ſame kind of ſeeds on different ſoils.

[1] Many people having expreſſed a deſire that i ſhould have plates of ſome of the profitable graſſes added to this piece, that moſt excellent man, the late Mr. Price of Foxley, whoſe extraordinary character i ſhall always revere, and do intend to give a ſketch of on ſome future occaſion, kindly condeſcended to employ his pencil, which in the opinion of the beſt judges was equal to things of a much ſuperior nature, in making me ſeveral drawings from the plants themſelves, and a very able hand has ſupplied the reſt and engraved them all.

[m] Thus Gmelin Flor. Lapp. mentions four of the meadow graſſes which he ſays have for a long time perplexed botaniſts of great reputation. And the editor of Ray's Synopſis, p. 402. doubts whether five graſſes which are put down as different by Petiver be not only varieties of a graſs mentioned before. I have many ſpecimens of this graſs in my collection differing in color, ſtature and outward aſpect, which yet moſt likely are of the ſame ſpecies.

Tab. 1.

Vernal Grass.

R. Rice delin.

OBSERVATIONS ON GRASSES.

A Table of English GRASSES.

GENUS 1.
VERNAL grass, *Tab.* 1. Anthoxanthum *Odoratum* *H. 10. R. 398. 1.

GENUS 2.
MAT grass   Nardus *Stricta* H. 20. R. 393.2.

GENUS 3.
Manured   CANARY grass Phalaris *Canariensis* H. 20. R. 394.
Sea   CANARY   Phalaris *Arenaria* H. 21. R. 398.4.
Reed   CANARY   Phalaris *Arundinacea* H. 21. R. 400. 1.
Ribband   CANARY   Phalaris *ibid. b. ibid.*

GENUS 4.
Green   PANIC grass   Panicum *Viride* H. 21. R. 393.1.
Loose   PANIC   Panicum *Crusgalli* H. 22. R. 394.2.
Cock's-foot PANIC   Panicum *Sanguinale* H. 22. R. 399. 2.
Creeping   PANIC   Panicum *Dactylon* H. 22. R. 399.1.

* N. B. H refers to the Flora Britannica of Mr. Hudson.

GENUS 5.

Meadow CAT's-TAIL grafs Phleum *Pratenſe* H. 22. R. 398.1.
Branched CAT's-TAIL Phleum *Paniculatum* H. 23.
Bulbous CAT's-TAIL Phleum *Nodoſum* H. 23. R. 398. 3.

GENUS 6.

Meadow FOX-TAIL grafs, *Tab.* 2. Alopecurus *Pratenſis* H. 23. R. 396. 1.
Field FOX-TAIL Alopecurus *Myoſuroides* H. 23. R. 397.
Bulbous FOX-TAIL Alopecurus *Bulboſus* H. 24. R. 397.3.
Flote FOX-TAIL Alopecurus *Geniculatus* H. 24. R. 396.2.

GENUS 7.

FEATHER grafs Stipa *Pennata* H. 24. R. 393.3.

GENUS 8.

Smooth COCK's-FOOT grafs Dactylis *Cynoſuroides* H. 25. R. 393.4.
Rough COCK's-FOOT Dactylis *Glomeratus* H. 25. R. 400. 2.

GENUS 9.

MILLET grafs Milium *Effuſum* H. 25. R. 402. 1.

Tab. 2.

Meadow Fox-tail Grass.

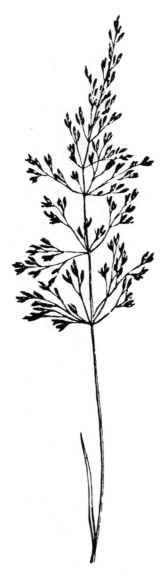

Tab. 2.

Fine Bent Grass.

R. Price delin.

GENUS 10.

Silky	BENT grafs	Agroftis *Spica venti*
	H. 26. R. 405. 17.	
Brown	BENT	Agroftis *Canina*
	H. 26.	
Red	BENT	Agroftis *Rubra*
	H. 26. R. 394. 4.	
Creeping	BENT	Agroftis *Stolonifera*
	H. 27. R. 402. 2.	
Marfh	BENT	Agroftis *Paluftris*
	H. 27. R. 404. 11.	
Fine	BENT *Tab.* 3.	Agroftis *Capillaris*
	H. 27. R. 402. 4.	
Wood	BENT	Agroftis *Sylvatica*
	H. 28. R. 404. 13.	
Small	BENT	Agroftis *Minima*
	H. 28. R. appendix.	

GENUS 11.

Crefted	HAIR grafs	Aira *Criftata*
	H. 28. R. 396. 3.	
Purple	HAIR	Aira *Cærulea*
	H. 29. R. 404. 8.	
Water	HAIR	Aira *Aquatica*
	H. 29. R. 402. 3.	
Turfy	HAIR	Aira *Cefpitofa*
	H. 29. R. 403. 5.	

Mountain

Mountain	HAIR	*Tab.* 4.	Aira *Flexuosa*
	H. 30. R. 407.8,9.		
Small leaved	HAIR		Aira *Setacea*
	H. 30.		
Grey	HAIR		Aira *Canescens*
	H. 30. R. 405. 16.		
Early	HAIR		Aira *Præcox*
	H. 31. R. 407.10.		
Silver	HAIR	*Tab.* 5.	Aira *Caryophillea*
	H. 31. R. 407.7.		

GENUS 12.

MELIC grass Melica *Nutans*
H. 31. R. 403.6.

GENUS 13.

Middle	QUAKING grass	Briza *Media*
	H. 32. R. 412.1.	
Small	QUAKING	Briza *Minor*
	H. 32. R. 412.2.	

GENUS 14.

Water	MEADOW grass	Poa *Aquatica*
	H. 32. R. 411. 13.	
Common	MEADOW	Poa *Trivialis*
	H. 33. R. 409.2.	
Great	MEADOW *Tab.* 6.	Poa *Pratensis*
	H. 33. R. 409. 3.	
Creeping	MEADOW	Poa *Compressa*
	H. 33. R. 409.5.	

Narrow-

Tab. 4.

Mountain Hair Grass.

R. Rice delin.

Tab. 5.

Silver Hair-Grass.

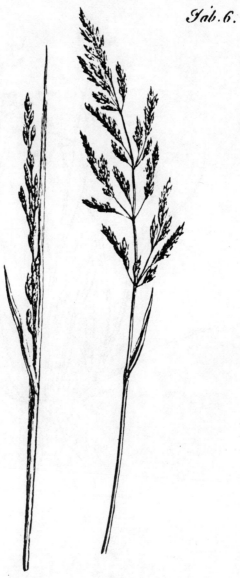

Tab. 6.

Great Meadow Grass.

Annual Meadow Grass

Tab. 8.

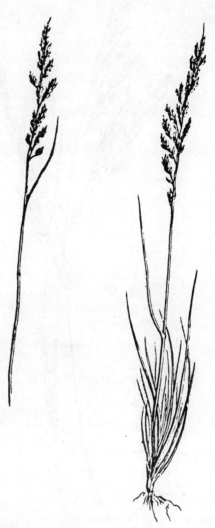

Sheeps Fescue Grass.

Purple Fescue Grass.

OBSERVATIONS ON GRASSES.

Narrow-leaved MEADOW   Poa *Angustifolia*
H. 34. R. 409. 4.
Bulbous MEADOW   Poa *Bulbosa*
H. 34. R. 411. 12.
Hair-leaved MEADOW   Poa *Setacea*
H. 34.
Annual MEADOW *Tab.* 7.   Poa *Annua*
H. 34. R. 408. 1.
Wood MEADOW   Poa *Nemoralis*
H. 34.
Sea MEADOW   Poa *Maritima*
H. 35. R. 410. 7.
Hard MEADOW   Poa *Rigida*
H. 35. R. 410. 8.
Spiked MEADOW   Poa *Loliacea*
H. 35. R. 395. 4.

GENUS 15.

Sheep's FESCUE grafs, *Tab.* 8. Festuca *Ovina*
H. 36. R. 410. 9.
FESCUE   Festuca *Vivipara*
ibid. *b.*
Hard FESCUE   Festuca *Duriuscula*
H. 36. R. 413. 4.
Purple FESCUE *Tab.* 9.   Festuca *Rubra*
H. 36.
Barren FESCUE   Festuca *Bromoides*
H. 37. R. 415. 13.

                                Wall

Wall	FESCUE	Festuca *Myurus*
	H. 37. R. 411.16.	
Tall	FESCUE	Festuca *Elatior*
	H. 37. R. 411.15.	
Small	FESCUE	Festuca *Decumbens*
	H. 38. R. 408.11.	
Flote	FESCUE Tab. 10.	Festuca *Fluitans*
	H. 38. R. 412.17.	
Spiked	FESCUE	Festuca *Loliacea*
	H. 38.	
Wood	FESCUE	Festuca *Sylvatica*
	H. 38. R. 394.	

GENUS 16.

Field	BROME grass	Bromus *Secalinus*
	H. 39. R. 413.5. 414.7,8.	
Corn	BROME	Bromus *Arvensis*
	H. 39. R. 414.9.	
Upright	BROME	Bromus *Erectus*
	H. 39.	
Wall	BROME	Bromus *Ciliatus*
	H. 40. R. 413.2.	
Barren	BROME	Bromus *Sterilis*
	H. 40. R. 412.1.	
Tall	BROME	Bromus *Giganteus*
	H. 40. R. 415.11.	
Wood	BROME	Bromus *Ramosus*
	H. 40. R. 415.10.	

Spiked

Flote Fescue Grass.

Spiked	BROME H. 41. R. 392.	Bromus *Pinnatus*

GENUS 17.

Naked	OAT grafs H. 41. R. 389. *b*.	Avena *Nuda*
Bearded	OAT H. 41. R. 389.7.	Avena *Fatua*
Meadow	OAT H. 42. R. 405.1.	Avena *Pratenfis*
Rough	OAT H. 42. R. 406.2.	Avena *Pubefcens*
Tall	OAT H. 42. R. 406.4.	Avena *Elatior*
Yellow	OAT H. 42. R. 407.5.	Avena *Flavefcens*

GENUS 18.

Common	REED grafs H. 43. R. 401.1.	Arundo *Phragmites*
Branched	REED H. 43. R. 401.2.	Arundo *Calamagroftis*
Small	REED H. 43. R. 401.3.	Arundo *Epigeios*
Sea	REED H. 43. R. 393.1.	Arundo *Arenaria*

GENUS 19.

Perennial	DARNEL grafs H. 44. R. 395. 2.	Lolium *Perenne*
		Annual

380 OBSERVATIONS ON GRASSES.

Annual DARNEL Lolium *Temulentum*
H. 44. R. 395.1.

GENUS 20
LYME grafs Elymus *Arenarius*
H. 44. R. 390.3.

GENUS 21.
Common WHEAT grafs Triticum *Repens*
H. 45. R. 390.1.
Bearded WHEAT Triticum *Caninum*
H. 45. R. 390.2.
Sea WHEAT Triticum *Junceum*
H. 45. R. 391.5.

GENUS 22.
BARLEY grafs Hordeum *Murinum*
H. 46. R. 392.3.

GENUS 23.
RYE grafs Secale *Villofum*
H. 46. R. 392.4.

GENUS 24.
Crefted DOG's-TAIL, grafs *Tab* 11. Cynofurus *Criftatus* H. 47. R. 398.2. 399.3.
Rough DOG's-TAIL Cynofurus *Echinatus*
H. 47. R. 397.5.
Blue DOG's-TAIL Cynofurus *Cæruleus*
H. 47. R. 399.4.
Bearded DOG's-TAIL Cynofurus *Paniceus*
H. 47. R. 396.4.

GENUS

Tab. 11

Crested Dogs tail Grass.

GENUS 25.

SOFT grafs   Holcus *Lanatus*
H.   R. 404. 14.

Genus the firft. VERNAL. *Tab.* 1.

This grafs grows very commonly on dry hills, and likewife on found rich meadow land. It is one of the earlieft graffes we have, and from its being found on fuch kinds of paftures as fheep are fond of, and from whence excellent mutton comes, it is moft likely to be a good grafs for fheep paftures. It gives a grateful odor to hay.

ADDITION. *This grafs i have found on all kinds of grounds, from the moft fandy and dry to the moft ftiff and moift, and even in bogs. It is very plentiful in the beft meadows about London, viz. towards Hampftead and Hendon. It is very eafy to gather, as i have found by experience; as it fheds its feeds upon the leaft rubbing.*

Genus the fixth. *Meadow* FOX-TAIL. *Tab.* 2.

This grafs as well as the foregoing is found in great plenty in our beft meadows about London, and i believe makes very good hay. Linnæus fays that it is a proper grafs to fow on grounds that have been drained.

ADDITION. *I am informed that the beft hay*

hay which comes to London is from the meadows, where this grass abounds. I saw this spring a meadow not far from Hampstead, which consisted of this grass chiefly with some of the vernal grass and the corn brome grass. This grass is scarce in many parts of England, particularly Herefordshire, Berkshire and Norfolk. It might be gathered at almost any time of the year from hay ricks, as it does not shed its seeds without rubbing, which is the case of but few grasses.

Water FOX-TAIL.

This is also found in the meadows about town, that are sound but lye under water in the winter, and perhaps might be proper to sow on such grounds.

Genus the tenth. *Marsh* BENT.

ADDITION. *This grass grows very commonly in moist grounds and ditches in many parts of England, where i have been. I shall say more of it under article Flote* FESCUE *grass.*

Fine BENT. *Tab.* 3.

This grass i have always found in great plenty on the best sheep pastures, as on Malvern hills, and on all the high grounds in Herefordshire, that are remarkable for good mutton.

ADDITION. *I may add on Bagshot heath, and the best sheep pastures in Berkshire, Oxfordshire, and Norfolk.*

Genus the eleventh. *Mountain* HAIR. *Tab.* 4.

The same may be said of this grass as of the foregoing. It grows in great plenty on Bagshot heath.

Silver HAIR. *Tab.* 5.

This also is found on the same kind of pasture as the two foregoing.

Genus the fourteenth. *Great* and *narrow-leaved* MEADOW. *Tab.* 6.

These grasses are common in our best meadow grounds, and i believe make good pasture and hay.

ADDITION. *I have found them frequently on banks by the road side, and near ditches, even where they were not to be found in the adjoyning meadows, and pastures.*

Annual MEADOW. *Tab.* 7.

This grass makes the finest of turfs. It grows every where by way sides, and on rich sound commons. It is called in some parts the Suffolk grass. I have seen whole fields of it in High Suffolk without any mixture of other grasses, and as some of the best salt butter we have in London comes from that county, it is most likely to be the best grass for the dairy. I have seen a whole park in Suffolk covered with this grass, but whether it affords good venison i cannot tell, having never tasted of any from it. I should rather think

think not, and that the best pasture for sheep is also the best for deer. However this wants trial. I remarked on Malvern hill something particular in relation to this grass. A walk that was made there for the convenience of the water drinkers, in less than a year was covered in many places with it, tho' i could not find one single plant of it besides in any part of the hill. This was no doubt, owing to the frequent treading, which above all things makes this grass florish, and therefore it is evident that rolling must be very serviceable to it.

ADDITION. *It has been objected that this grass is not free from bents, by which word is meant the flowering stems. I answer that this is most certainly true, and that there is no grass without them. But the flowers and stems do not grow so soon brown as those of other grasses, and being much shorter they do not cover the radical leaves so much, and therefore this grass affords a more agreeable turf without mowing, than any other whatever that i know of.*

Sheeps FESCUE. *Tab.* 8.

This is the grass so much esteemed in Sweden for sheep.

Gmelin. Flor. Sibir. says that the Tartars choose to fix during the summer in those places where there is the greatest plenty of this grass; because

because it affords a most wholesome nourishment to all kinds of cattle, but chiefly sheep; and he observes that the sepulchral monuments of the antient Tartars are mostly found in places that abound with this grass, which shews, adds he, that it has long been valued amongst them.

I have among my grasses a specimen of it, but do not remember where I found it. I am certain it is not common in any of the places where I have been. Perhaps upon examination it may be found on places famous for our best mutton, as Banstead Downs, Church-Stretton in Shropshire, some parts of Wales, &c.

ADDITION. *I have since found this grass in great plenty in many parts of England and Wales; indeed on all the finest sheep pastures in Herefordshire, Berkshire, Oxfordshire, Norfolk, &c. The reason why I thought it not common, was, that it is an early grass, and had shed its seeds, before I usually made my searches in those places where it only grows. I must also observe that, contrary to what Linnæus says, either the sheep or some other animals do eat the flowering stems of this grass, for upon Banstead Downs there was nothing to be seen but the radical leaves of it, unless amongst the bushes near the hedges, where it was guarded from the sheep.*

Genus the fifteenth. *Purple* FESCUE. *Tab.* 9.

ADDITION. *This grafs i have always found along with the fine* BENT *and silver* HAIR-GRASS, *particularly on Banstead Downs in great plenty in a place inclosed in order to keep the sheep out. From hence i am inclined to think that this is the chief grafs all over the Downs, but as the flowering stems in the other parts were intirely gone, unlefs along the hedges, I could not be certain.*

Flote FESCUE. *Tab.* 10.

I have no knowledge of the qualities of this grafs from my own experience, but shall quote something concerning it out of a piece published in the Amœn. Academ. vol. 3. entitled Plantæ Efculentæ. The author fays there, artic. 90. that the feeds of this grafs are gathered yearly in Poland, and from thence carried into Germany and sometimes into Sweden, and fold under the name of manna feeds. Thefe are much ufed at the tables of the great on account of their nourifhing quality and agreeable tafte. It is wonderfull, adds the author, that amongft us thefe feeds have hitherto been neglected, fince they are fo eafily collected and cleanfed.

ADDITION. *Mr. Dean, a very fenfible farmer at Rufcomb, Berkfhire, affured me that a field always lying under water of about four acres,*

that

that was occupied by his father when he was a boy, was covered with a kind of grass, that maintained five farm-horses in good heart from April to the end of harvest, without giving them any other kind of food, and that it yielded more than they could eat. He at my desire brought me some of the grass, which proved to be the flote FESCUE with a mixture of the marsh BENT; whether this last contributes much towards furnishing so good pasture for horses i cannot say. They both throw out roots at the joynts of the stalks, and therefore likely to grow to a great length. In the index of dubious plants at the end of Ray's Synopsis, there is mention made of a grass under the name of Gramen caninum supinum longissimum, growing not far from Salisbury 24 feet long. This must by its length be a grass with a creeping stalk; and that there is a grass in Wiltshire growing in watery meadows so valuable, that an acre of it lets from 10 to 12 pounds, i have been informed by several persons. These circumstances incline me to think it must be the flote fescue; but whatever grass it be, it certainly must deserve to be inquired after.

There is a clamminess on the ear of the flote fescue, when the seeds are ripe that tastes like honey, as i have often found, and for this reason perhaps they are called manna seeds.

Linnæus

Linnæus Flor. Suec. art. 95. *says, that the bran of this grass will cure horses troubled with bots, if kept from drinking for some hours.*

Genus the seventeenth. *Yellow* OAT.

This grass is found in great plenty in some grounds where the sheep's FESCUE, *the fine* BENT, *and the crested* DOG-TAIL, *grow, and therefore likely to be good for sheep. It is also not uncommon in good meadows.*

Genus the nineteenth. *Perennial* DARNEL.

This grass is well known, and cultivated all over England; and it is to be hoped the success we have had with it will in time encourage our farmers to take the same pains about some others that are no less valuable, and are full as easy to be separated. It makes a most excellent turf on found rich land where it will remain.

If i may judge by the venison i have eat out of a paddock, that was chiefly filled with this grass, i would by no means recommend it for parks. I know it will be said that venison is never good out of a paddock, that the deer must have room to range, trees to browse on, &c. I grant there is some reason for saying this, but i believe in general it is more owing to want of proper food, viz. good grass, than merely to confinement; for paddocks are generally made

by

OBSERVATIONS ON GRASSES.

by converting some rich spot near the house that has constantly been manured, and of course is full of grasses fitter for the dairy or the stable than for deer, which hardly ever is the case of large parks. No man will, i suppose, pretend to make good pork from a hog fed with grains instead of peas, tho' he has the liberty of choosing as much ground as he pleases, and where he pleases.

This grass is called in many counties *rye* grass. It were to be wished that the old name might prevail, because there is a genus of grass, viz. the 22d. known by the name of *rye* all over the kingdom, of which genus there is a species that ought to bear the same generical name.

ADDITION. *I have since eaten venison out of a large park, where there was much of this grass, and it was no better than that out of the paddock. I should be apt to think from hence that this grass would not be proper for sheep, as i have always observed that the same kind of ground which yields good venison yields also good mutton. For what particular uses it is good, wants to be tryed, whether for the dairy, for fatting cattle, or for horses. Many are tempted by the facility of procuring the seed of this grass to lay down grounds near their houses, where they want to have a fine*
turf

turf with it; for which purpose unless the soil be very rich a worse grass cannot be sown, as it will certainly die off in a very few years intirely.

Genus the twenty fourth. Crested DOG-TAIL. *Tab.* 11.

This grass i imagine is proper for parks. I know one where this abounds, that is famous for excellent venison. It may perhaps be as good for sheep.

ADDITION. *That it is good for sheep i have since found by experience. The best mutton i have tasted, next to that which comes from hills where the purple and sheep's fescue, the fine bent, and the silver hair grasses abound, having been from sheep fed with it.*

It makes a very fine turf upon dry sandy or chalky soils, as i have seen in many parts of Berkshire, but unless swept over with the scythe, its flowering stems will look brown; which is the case of all grasses, which are not fed by variety of animals. For that some animals will eat the flowering stems is evident by commons, where scarcely any parts of grasses appear but the radical leaves.

Order of coming into ear of the above mentioned grasses.

Annual MEADOW
Meadow FOX-TAIL

VERNAL

VERNAL

Great MEADOW
Narrow-leaved MEADOW
Crested DOG-TAIL
Sheep's FESCUE
Purple FESCUE
Fine BENT
Marsh BENT
Silver HAIR
Yellow OAT
Flote FESCUE

The whole time from the beginning of May till about the middle of June.

Ει—κεν και σμικρον επι σμικρῳ καταθειο
Και θαμα τυθ' ερδοις, ταχα κεν μεγα και το γενοιο.
Hesiod.

FINIS.

HISTORY OF ECOLOGY
An Arno Press Collection

Abbe, Cleveland. **A First Report on the Relations Between Climates and Crops.** 1905

Adams, Charles C. **Guide to the Study of Animal Ecology.** 1913

American Plant Ecology, 1897-1917. 1977

Browne, Charles A[lbert]. **A Source Book of Agricultural Chemistry.** 1944

Buffon, [Georges-Louis Leclerc]. **Selections from Natural History, General and Particular, 1780-1785.** Two volumes. 1977

Chapman, Royal N. **Animal Ecology.** 1931

Clements, Frederic E[dward], John E. Weaver and Herbert C. Hanson. **Plant Competition.** 1929

Clements, Frederic Edward. **Research Methods in Ecology.** 1905

Conard, Henry S. **The Background of Plant Ecology.** 1951

Derham, W[illiam]. **Physico-Theology.** 1716

Drude, Oscar. **Handbuch der Pflanzengeographie.** 1890

Early Marine Ecology. 1977

Ecological Investigations of Stephen Alfred Forbes. 1977

Ecological Phytogeography in the Nineteenth Century. 1977

Ecological Studies on Insect Parasitism. 1977

Espinas, Alfred [Victor]. **Des Sociétés Animales.** 1878

Fernow, B[ernhard] E., M. W. Harrington, Cleveland Abbe and George E. Curtis. **Forest Influences.** 1893

Forbes, Edw[ard] and Robert Godwin-Austen. **The Natural History of the European Seas.** 1859

Forbush, Edward H[owe] and Charles H. Fernald. **The Gypsy Moth.** 1896

Forel, F[rançois] A[lphonse]. **La Faune Profonde Des Lacs Suisses.** 1884

Forel, F[rançois] A[lphonse]. **Handbuch der Seenkunde.** 1901

Henfrey, Arthur. **The Vegetation of Europe, Its Conditions and Causes.** 1852

Herrick, Francis Hobart. **Natural History of the American Lobster.** 1911

History of American Ecology. 1977

Howard, L[eland] O[ssian] and W[illiam] F. Fiske. **The Importation into the United States of the Parasites of the Gipsy Moth and the Brown-Tail Moth.** 1911

Humboldt, Al[exander von] and A[imé] Bonpland. **Essai sur la Géographie des Plantes.** 1807

Johnstone, James. **Conditions of Life in the Sea.** 1908

Judd, Sylvester D. **Birds of a Maryland Farm.** 1902

Kofoid, C[harles] A. **The Plankton of the Illinois River, 1894-1899.** 1903

Leeuwenhoek, Antony van. **The Select Works of Antony van Leeuwenhoek.** 1798-99/1807

Limnology in Wisconsin. 1977

Linnaeus, Carl. **Miscellaneous Tracts Relating to Natural History, Husbandry and Physick.** 1762

Linnaeus, Carl. **Select Dissertations from the Amoenitates Academicae.** 1781

Meyen, F[ranz] J[ulius] F. **Outlines of the Geography of Plants.** 1846

Mills, Harlow B. **A Century of Biological Research.** 1958

Müller, Hermann. **The Fertilisation of Flowers.** 1883

Murray, John. **Selections from *Report on the Scientific Results of the Voyage of H.M.S. Challenger During the Years 1872-76.*** 1895

Murray, John and Laurence Pullar. **Bathymetrical Survey of the Scottish Fresh-Water Lochs.** Volume one. 1910

Packard, A[lpheus] S. **The Cave Fauna of North America.** 1888

Pearl, Raymond. **The Biology of Population Growth.** 1925

Phytopathological Classics of the Eighteenth Century. 1977

Phytopathological Classics of the Nineteenth Century. 1977

Pound, Roscoe and Frederic E. Clements. **The Phytogeography of Nebraska.** 1900

Raunkiaer, Christen. **The Life Forms of Plants and Statistical Plant Geography.** 1934

Ray, John. **The Wisdom of God Manifested in the Works of the Creation.** 1717

Réaumur, René Antoine Ferchault de. **The Natural History of Ants.** 1926

Semper, Karl. **Animal Life As Affected by the Natural Conditions of Existence.** 1881

Shelford, Victor E. **Animal Communities in Temperate America.** 1937

Warming Eug[enius]. **Oecology of Plants.** 1909

Watson, Hewett Cottrell. **Selections from *Cybele Britannica.*** 1847/1859

Whetzel, Herbert Hice. **An Outline of the History of Phytopathology.** 1918

Whittaker, Robert H. **Classification of Natural Communities.** 1962